Research in Human Development
Volume 1, Number 3

Special Issue: Contextual Influences on Life Span/Life Course Development
Guest Editor: Jacquelynne S. Eccles

Lifespan Psychology: From Developmental Contextualism
to Developmental Biocultural Co-constructivism   123
  Paul B. Baltes and Jacqui Smith

The Interpenetration of Culture and Biology
in Human Development   145
  Cynthia García Coll

Whose Lives? How History, Societies, and Institutions Define
and Shape Life Courses   161
  Karl Ulrich Mayer

Civic Engagement, Political Identity, and Generation
in Developmental Context   189
  Abigail J. Stewart and Christa McDermott

How Gene–Environment Interactions Shape Biobehavioral
Development: Lessons From Studies With Rhesus Monkeys   205
  Stephen J. Suomi

Lifespan Psychology: From Developmental Contextualism to Developmental Biocultural Co-constructivism

Paul B. Baltes and Jacqui Smith
Max Planck Institute for Human Development

Lifespan psychology has always been associated with a family of scripts about development and aging. An initial set of scripts included proposals about developmental contextualism at the macro-level (e.g., age-graded, history-graded, and nonnormative influences). Recent theoretical efforts to link evolutionary and ontogenetic perspectives engendered an additional set of interrelated scripts about the nature and consequences of human development. Proposals about the biocultural architecture of the lifespan highlight its inherent incompleteness and aging-based increase in incompleteness and vulnerability. Age-related differences in the overall allocation of resources (from growth to maintenance and the regulation of loss) as well as the general-purpose mechanisms of selection, optimization, and compensation orchestrate adaptive development and aging within the constraints of the biocultural architecture. We argue that this package of conceptions converges with the notion of developmental biocultural co-constructivism and specifies the zone within which human development can be expressed.

Without downgrading the role of alternative theoretical endeavors and their powerful impact on the developmental sciences (see Elder, 1998; Lerner, 2002; and Magnusson, 1996, for reviews), lifespan researchers like to argue that their theoretical orientations have considerably enriched, if not transformed, the field of developmental psychology (e.g., P. B. Baltes, 1987, 1997; P. B. Baltes, Lindenberger, & Staudinger, 1998; Staudinger & Lindenberger, 2003). In essence, the lifespan orientation was designed not only to highlight that development continues beyond childhood and adolescence but also to bring to the foreground additional content phenomena and principles of determination. When researchers

---

Requests for reprints should be sent to Paul B. Baltes or Jacqui Smith, Center for Lifespan Psychology, Max Planck Institute for Human Development, Lentzeallee 94, 14195 Berlin, Germany. E-mail: sekbaltes@mpib-berlin.mpg.de or smith@mpib-berlin.mpg.de

view development as being lifelong rather than as restricted to a single age period, topics such as wisdom, intergenerational dynamics, and the influence of changing historical contexts on individual development spring to mind. Consider, for instance, the changes in the directional influence and power of reciprocal socialization when contrasting parents and infants with the counterpart situation in the second half of life, parents and their adult children (Hetherington & Baltes, 1988). Or consider the consequences of the historical increase in average life expectancy, from about 45 years in 1900 to about 80 years in the year 2000. Such dramatic focal changes on social transactions or history-conditioned phenomena are more difficult to identify when the theoretical lens is set for only a single age period, such as childhood.

Historically, there has been a long history of lifespan thinking. One major source dates back to Tetens (1777), who published a monumental work on human development more than 200 years ago. This work explored functional changes in human characteristics across the lifespan and drew attention to the fundamental questions of modifiability of development, including the impact of historical changes (P. B. Baltes et al., 1998; Lindenberger & Baltes, 2000). There have been equally strong voices from the more recent past in North America. In her presidential address to American Psychological Association Division 20 in 1958, for instance, the noted child developmentalist Nancy Bayley captured the sentiments of the mid-20th century: "Psychological theory and research will benefit in many ways if the research is planned, carried out, and interpreted within the frame of reference of the lifespan and the continuous processes of change that characterize all behavior" (Bayley, 1963, p. 137). The West Virginia Conference Series on lifespan developmental psychology, first organized in 1969, was a collective effort to articulate lifespan theory and research and to promote lifespan perspectives within developmental psychology (e.g., Goulet & Baltes, 1970). Although much has been achieved since then in terms of theoretical and methodological advances, there still remain many lacunae and areas to be elaborated.

As an effort toward further elaboration, this article focuses on lifespan proposals and research about the multilevel systems of contextual influences on development and the evolutionary and ontogenetic design of these influences. We begin with a summary of the scripts about developmental contextualism that characterized the beginnings of lifespan psychology. At that time, and in close collaboration with life course sociologists, three classes of interactive contextual biosocial influences were identified: (a) age-graded, (b) history-graded, and (c) non-normative (e.g., P. B. Baltes, Reese, & Lipsitt, 1980). Subsequently, and in line with recent theoretical efforts to refine conceptions of developmental and biocultural contextualism (e.g., P. B. Baltes, 1997; P. B. Baltes & Singer, 2001; S.-C. Li, 2003), a new set of lifespan scripts have evolved. These scripts, which are outlined in later sections of this article, deal with: (a) the overall "architectural" structure of ontogeny and the dynamics between biological and cultural factors; (b) the differential allocation of resources across the lifespan, from a primary emphasis on growth to maintenance and regulation of loss; and (c) the systemic operation of a set of mechanisms (selection, optimization, compensation) that orchestrate adaptive (successful) development and aging.

We conclude with a consideration of the implications of these scripts for the last phase of life. Because the lens of contemporary lifespan development extends far beyond what was possible in earlier times (e.g., to research on the oldest-old and centenarians; Smith, 2001), a new sociodemographic scenario is emerging: that of the graying of the population. In our view (P. B. Baltes & Smith, 1999, 2003), this novel scenario promises to generate a new situation of history-graded biocultural contextualism. In this last phase of life, psychological functioning may be characterized by aspects of change and constellations of factors that are quite distinct from the causal and processual network that is operative at earlier phases of life.

LIFESPAN SCRIPTS ABOUT THE MACRO-CONTEXTS OF DEVELOPMENT

We begin with an overview of the early efforts of lifespan theorists to identify frameworks for considering contextualism at a macro-level of analysis and the impact of such contexts on the production of commonalities and differences in development (P. B. Baltes et al., 1980). It is important to note that these efforts are consistent with our current view of contextualism (i.e., biocultural contextualism as co-constructivism) in that they refer to the ways in which intrinsic and extrinsic factors are structured across the lifespan. Specifically, the initial lifespan scripts proposed that biological and environmental contexts of development are structured at multiple levels across the lifespan by three classes of influences: (a) age-graded (ontogenetic), (b) history-graded, and (c) non-normative (idiosyncratic) influences (see also P. B. Baltes, Cornelius, & Nesselroade, 1979).

As a whole, the cumulative interactions and co-productions of these classes of contextual influences contribute, on the one hand, to much commonality and continuity in the nature of developmental change. On the other hand, they contribute to interindividual and subgroup differences in status and to differences in the direction and level of intraindividual change over time. For instance, age-graded influences can vary systematically by social class, gender, cohort, or ethnicity. Incidentally, a failure to recognize the inherent individual and group differentiation associated with the three classes of influences was a major source of misunderstanding between sociologists and psychologists (P. B. Baltes & Nesselroade, 1984; Dannefer, 1992; Mayer, 2003).

This macro-view of intrinsic and extrinsic developmental contexts highlighted the concept of plasticity as a fundamental lifespan script (P. B. Baltes & Schaie,

1976; Gollin, 1981; Lerner, 1984). *Plasticity* was defined as the range of human development that was possible under varying constellations of age-graded, history-graded, and non-normative influences. It can be studied by means of experimental simulations of development in which different learning histories are examined, for instance, by time-compressed designs of cognitive training or methods of testing the limits (P. B. Baltes, Reese, & Nesselroade, 1977; Kliegl, Smith, & Baltes, 1989; Lindenberger & Baltes, 1995). Although lifespan researchers typically argue for the need for complex longitudinal designs and for a creative combination of descriptive with explanatory longitudinal research (e.g., Kruse, Lindenberger, & Baltes, 1993; Magnusson & Casaer, 1993; Schaie, 1965), it is part of the pragmatics of science that natural time takes too long for such idealized designs to be realizable within the active lifetime of a given researcher or research team.

The Multilevel Organization of Context in Lifespan Development

Because of the complexity and plasticity of the conditions shaping the course of human development, the general approach of lifespan psychologists has always been to highlight the *pluralistic* (multidimensional, multilevel), *transactional*, and *dynamic nature of contextual influences* on individual change. Indeed, the course of individual development itself is considered as a changing phenomenon (see also Elder, 1998; Mayer, 2003; Riley, Foner, & Riley, 1999). Thus, when P. B. Baltes and his colleagues (P. B. Baltes et al., 1980; P. B. Baltes et al., 1977) distinguished among three sets of contextual influences—normative age-graded, normative history-graded, and non-normative—their intention was to alert researchers to consider multiple levels of explanation for observed age-related and individual differences and change over time.

Normative age-graded (ontogenetic) influences. Nowadays, few would dispute that age-related factors play a pervasive, cumulative organizational role in the structuring of the biological and environmental contexts of development over the entire lifespan. *Normative* is used here in the statistical sense to indicate that sets of events occur in highly similar ways (timing and duration) for the majority of individuals in a given society or subculture.

There is a long tradition of examining age-graded contexts and systems of influence in the first 20 years of the human lifespan. More conceptual effort is needed, however, to specify the mechanisms and nature of age-graded biological and environmental contexts of developmental change in the period of adulthood and old age. Consider first our knowledge of age-graded biological influences in adulthood: Compared to the first 20 years of life, when much is known about the normative correlations between chronological age and aspects of biological maturity, we have relatively less-detailed knowledge about temporal sequences and age–biology associations in mid-adulthood and old age. Regarding the second half of life, the value of searching for biological "age markers" is disputed (e.g., McClearn, 1997) and theorists suggest that, although some organs and biological systems show regular functional changes over time, there is no general "program" of aging as such: Stochastic processes contribute to increased interindividual variation with age after young adulthood (e.g., Finch & Kirkwood, 2000; Kirkwood, 2003).

The specification of normative age-graded socialization events, developmental tasks, and ecologies is also a research area ripe for contemporary update (e.g., Havighurst, 1972; Neugarten, 1969; Settersten, 1999). Several recent sociologically oriented reviews suggest that social expectations, ecologies of development, and pathways of age-graded transitions across the life course may be changing (e.g., Ferraro, 2001; Heckhausen & Dweck, 1998; Mayer, 2003; Sampson, Morenoff, & Gannon-Rowley, 2002; Shanahan, 2000). There is much debate whether contemporary societal and environmental contexts have become more or less age-structured and whether variation (diversity) within age stratas and age cohorts has increased (Dannefer & Uhlenberg, 1999; Mayer, 2003; Settersten, 1999).

History-graded influences. History-graded influences also involve biological and environmental contexts and contribute both to short- and long-term changes in developmental trajectories that may differentiate cohorts. From a discipline viewpoint, this is the territory of historians and sociologists (Elder, 1998; Mayer, 2003; Riley et al., 1999). However, insights from developmental biologists would help complete the picture of the biocultural dimension of history-graded influences.

Examples of history-graded influences are economic depression, war, social revolution, major epidemics, technological advances, major educational changes, changes in demographic structure and modernization, and changes in the content and practices of nutrition and other forms of health behavior. Research on birth cohort effects originally made the perhaps strongest case for consideration of historical contextualism (Elder, 1998). More recently, the lens has been extended to larger time dimensions and efforts to specify the causal and processual particulars. In the last 100 years, for instance, modernization has been associated with increasing variation in pathways to adult roles (e.g., Modell & Elder, 2002). Furthermore, cohorts are thought to differ in the level and shape of age trajectories on a wide array of dimensions, including intelligence (Flynn, 1999; Schaie, 1996), morbidity, and longevity (Maier & Vaupel, 2003; Vaupel et al., 1998).

Research on cohort differences is often used to support arguments that historical change brings improvement and progress. In this tradition, work on cognitive performance in late adulthood from studies in Sweden (e.g., Bäckman, Small, Wahlin, & Larsson, 2000) and Schaie's (1996) Seattle Longitudinal Study suggests that today's 70-year-olds are comparable to 65-year-olds who lived 30 years ago (see also Helmuth, 2003). The functional health of older adults also has improved (Manton, Stallard, & Corder, 1997). However, historical change may have negative consequences as well. For example, despite general positive shifts in cohort competencies of the intellect, Schaie (1996) reported negative cohort differences in performance on tasks assessing numerical ability: Younger birth cohorts exhibit lower numerical proficiency than older cohorts. Elbert (2003) suggested there may be negative consequences on the functional architecture of the brain associated with wars and other forms of violence or drug-related epidemics. Research on history-graded influences on the nature of individual development is reaching new heights and thereby strengthening proposals about the biocultural co-construction of ontogeny (P. B. Baltes & Singer, 2001; S.-C. Li, 2003).

Non-normative influences. The third set of influences reflects the unique individual–idiosyncratic biological and environmental events that are not clearly tied to ontogenetic or historical time. Typically, but not necessarily, they are statistically infrequent in a population and have no universal temporal and spatial sequence, yet they can have significant influences on the development of an individual (e.g., Bandura, 1982; Brim & Ryff, 1980). Examples include winning a lottery, chance personal encounters, career changes, relocation, serious accidents or illness, extended unemployment, divorce, unexpected death of significant others, migration, and being a victim of serious crime or warlike conditions.

The impact of non-normative events is thought to be especially powerful because such events disrupt the sequence and rhythm of the expected life cycle and so generate conditions of uncertainty (e.g., Diehl, 1999; Wrosch & Freund, 2001). Some, but not all, of these conditions are only minimally amenable to personal or social control and to long-term modification and therefore represent extreme situations of challenge. In individuals' life narratives, such events can be perceived as critical "turning points" (e.g., McAdams, 2001). The extent to which a non-normative life event will have long-lasting implications for life change likely depends on when it occurs across the life course and what type of change the event entailed in terms of social roles, functional status, and sense of identity. In our assessment, research on non-normative life events has been especially powerful if the focus was on the operation of multiple or conjoint life events and life situations in which developmental reserves were overtaxed or tested at limits (P. B. Baltes et al., 1998; Staudinger, Marsiske, & Baltes, 1993).

Changing Salience of Contextual Influences Over the Lifespan

Together, these three closely intertwined and co-constructive systems of influence, mediated through the developing individual and institutional structures and networks, have a cumulative effect producing regularities and individual differences in life pathways. None of these classes of biologically and environmentally based influences operates independently from the other. Such a focus on contextualism makes explicit the lack of full predictability of human development as well as the boundedness that individuals experience as they engage in efforts to construct and manage their lives (e.g., Brandtstädter & Lerner, 1999).

The combinational profile of the effects of various types of contextual influences may also differ by age (or historical) period. In this vein, P. B. Baltes et al. (1980) speculated about the relative salience of age-graded, history-graded, and non-normative influences at varying points in the lifespan. They hypothesized that age-graded influences are primarily important in child development and, perhaps to a lesser extent, in the transition to old age, whereas history-graded and non-normative influences become an increasingly dominant force of influence from young adulthood onward. The primacy of age-graded influences in childhood has long been supported by nomothetic developmental functions in such domains as cognitive and physical growth that are rather robust across cultures and historical time. The notion that age-graded influences might weaken with age, especially beyond the chronological age of average life expectancy, is consistent with evolutionary-based biological theories of aging (e.g., P. B. Baltes, 1997; Kirkwood, 2002). The reasoning is that evolutionary-based genetic control over the postreproductive phase of life has not been selected. It is also consistent with sociological theories that point to the relative absence of social roles for the majority of older adults in a population and the comparative insignificance of older adults post-retirement from the workforce for the organization of society (Rosow, 1985; Uhlenberg, 1988).

Evidence on the relative impact of normative history-graded influences over the entire life course was scarce in 1980 and is still limited. The effect of the timing of historical events in the lives of individuals most probably depends on the type of event, the extent to which it represents situations of gain or loss for individuals at different ages both in the short and long term, and the capacity of individuals to change at different ages (e.g., Elder, 1998; Wrosch & Freund, 2001).

P. B. Baltes et al. (1980) proposed that non-normative events take on an increasing salience in determining development after early adulthood. In part, this proposal was linked to the idea that the organizing role of age-graded biological and environmental factors declines in old age. Furthermore, it is likely that age-related losses in developmental reserve capacity play a crucial role. Together with the shifting valence of contextual influences across the lifespan, the proposal that non-normative influences gain in salience is in accord with findings that change in late adulthood is associated with losses in controllability, reduced potential to recover, and increased constraints on the possibilities of adopting alternative life pathways or compensatory measures (P. B. Baltes & Smith, 2003; Heckhausen, Dixon, & Baltes, 1989; Smith, 2003).

A LIFESPAN SCRIPT ABOUT BIOLOGY–CULTURE DYNAMICS

During recent years, we have developed a new set of lifespan scripts to make explicit the "causal" dynamics of lifespan development and strengthen our insights into the mechanisms of biocultural co-construction (e.g., P. B. Baltes, 1997). To begin, we describe an overarching script about the architecture of ontogeny, the landscape of human development. This overarching framework links basic principles of developmental biology to proposals about mechanisms and contexts of psychological development and aging and specifies the general forms and directional outcome of the linkage over time. Figure 1 summarizes the three central principles of this overarching framework. These principles, we argue, need to be considered as we attempt to understand the interactive system of age-graded, history-graded, and non-normative influences.

First, as depicted in the left panel of Figure 1, it is proposed that biological plasticity and genetic fidelity decrease as individuals reach the higher ages of the life course. This lifespan trajectory reflects the fact that biological evolution was oriented not toward optimizing old age but rather to optimizing reproductive fitness in early adulthood (e.g., Finch, 1990). As a consequence, the human genome in older age groups is more likely to be characterized by deleterious genetic expressions and reduced genetic fidelity. The outcome: Biogenetic plasticity decreases with age, although it continues to operate.

The second principle (middle panel of Figure 1) asserts that for human development to extend into the higher ages, new steps in the level and kind of cultural evolution and cultural resources are essential. To extend average life expectancy,

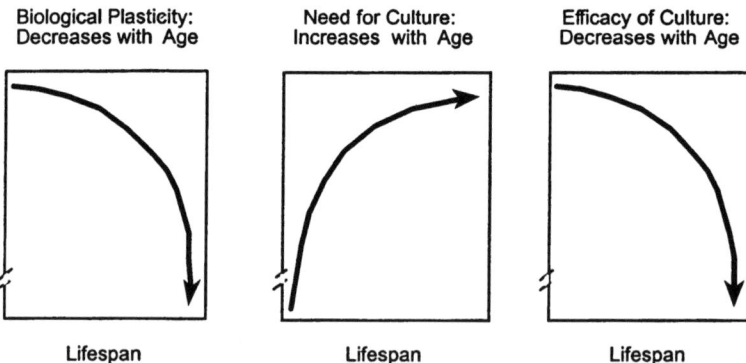

FIGURE 1 Biocultural architecture of life course: Schematic representation of three meta-principles that co-regulate human ontogeny. Together, these principles describe the dynamics between biology and culture across the life course that characterize an aging-associated increase in the incompleteness and vulnerability of individuals and populations as they age (modified from P. B. Baltes, 1997).

for instance, it takes more and more culture-based resources and practice to exploit the biogenetic potential that is inherent in the human genome. Thus, the material, technological, psychological, and social aspects of cultural evolution, not genetic evolution, are the driving force in recent quantitative and qualitative expansions of the life course (see also Durham, 1991).

The dilemma of modern times is in the lifespan function shown in the right panel of Figure 1. The efficacy of culture to exploit the genome and to compensate, if necessary, for the biological losses associated with aging decreases toward the end of the life course. The older the individual, the less improvement or repair is achieved given the same cultural input or intervention. In old age, for instance, it takes much more time and practice to reach the same cognitive output. Moreover, as individuals reach asymptotic performance in functions, further improvement is more difficult to achieve (P. B. Baltes et al., 1998).

This triangulated theoretical script of age-associated change in the biocultural architecture of the life course should be kept in mind when it comes to speculations about the future of adult development and aging in a population where more and more individuals will reach advanced old age. Of course, the script shown in Figure 1 characterizes a dynamic and evolving framework, and new science may change the constellation. For instance, the curves of three trajectories may be elongated to higher ages on the *x*-axis. Nevertheless, the direction of the age-related change, especially in the Fourth Age, will reflect the biocultural incompleteness of the architecture and its associated vulnerability and reduced potential.

A LIFESPAN SCRIPT ABOUT THE ALLOCATION OF RESOURCES: FROM GROWTH TO MAINTENANCE AND REGULATION OF LOSS

Another lifespan script proposed in recent years complements the overarching proposals about the biocultural dynamics and the operation of lifespan contexts (P. B. Baltes, 1997; Staudinger et al., 1993). This script outlines changes across the lifespan in the systemic configuration of three general functions of development: (a) growth; (b) maintenance, including repair and recovery; and (c) regulation of loss. It suggests that with increasing age individuals need to invest more and more of their internal and external resources into maintenance and management of loss as opposed to growth in order to assure adaptive efficacy and success. This systemic change sets boundary conditions for the operation and outcomes of developmental contexts.

With the phrase *adaptive function of growth* we refer to behaviors involved in reaching higher levels of functioning or adaptive capacity. Under the heading of *maintenance* we classify behaviors that ensure stability in levels of functioning in the face of a new contextual challenge or a loss in potential. Finally, regarding *regulation or management of loss*, we mean behaviors that organize functioning at lower levels when maintenance or recovery is no longer possible. In childhood, resources are primarily allocated to growth. During adulthood, the predominant allocation is toward maintenance. In old age, more and more resources are directed toward regulation or management of loss. Such a characterization of the lifespan, of course, is an oversimplification, because individual, functional (domain), contextual, and historical differences need to be taken into account. This lifespan script is about relative probability and prevalence.

The lifespan trajectories of resource investment into growth, maintenance, and regulation of loss have implications for the dynamics involved in the systemic and integrative coordination of these three functions. In this regard, it is not surprising that researchers of adult development have strong interests in topics such as goals and selection among goals as well as compensation for losses and the seemingly counterintuitive idea that conditions of deficit can breed advances through innovative efforts (P. B. Baltes et al., 1998; Cantor & Fleeson, 1994; Dixon & Bäckman, 1995; Eccles & Wigfield, 2002; Freund & Baltes, 2002; Uttal & Perlmutter, 1989). Such perspectives have led us to work on a general theory of adaptive development and the management of gains and losses, which we describe next.

A LIFESPAN SCRIPT OF ADAPTIVE DEVELOPMENT AND AGING: ORCHESTRATING SELECTION, OPTIMIZATION, AND COMPENSATION

During the last 10 years, we have worked on a general theory of adaptive (successful) development and aging (selective optimization with compensation [SOC]; P. B. Baltes, 1997; P. B. Baltes & Baltes, 1990; P. B. Baltes, Freund, & Li, in press; Freund & Baltes, 2002) that is consistent with the general scripts of lifespan development we have outlined. This is not the only theory that would fit this overall frame; however, it is a theory that was explicitly developed to suit this purpose.

Basic Framework of SOC

The theory was originally developed to describe successful aging and was called *selective optimization with compensation* (P. B. Baltes & Baltes, 1990). It involved proposals about the operation and coordination of three components: (a) selection of goals or outcomes, (b) optimization of means to reach these goals, and (c) compensation through the use of substantive means. These SOC components were subsequently construed as a general-purpose mechanism of development and adaptive functioning across the lifespan (P. B. Baltes, 1997). There are other similar approaches, most notably those of Brandtstädter (1998), Heckhausen and Schulz (1995), and Carstensen (1995).

Selection at the most general level refers to a development-enhancing process that in developmental biology is called *canalization* (Waddington, 1966). This selection process refers to a specification and narrowing down of a range of alternative outcome-oriented pathways that the scope of biocultural plasticity would permit in principle. It is a prerequisite for advances. However, selection may also be necessary when resources such as time, energy, and capacity are limited. To accommodate these two instantiations of selection and their different connotations, two forms of selection—elective and loss-based—have been differentiated (Freund & Baltes, 2002).

Optimization in the general sense refers to the acquisition, application, coordination, and maintenance of internal and external resources (means) involved in attaining higher levels of functioning. The relevant means are many, ranging from genetic expressions to health behavior, practice, cognitive skills, social support, education, and cognitive status.

Compensation, like optimization, refers to means; however, *compensatory means* serve to counteract losses in specific means previously used for goal attainment by using alternative (substitutive) means to maintain functioning. One example of compensation is the use of hearing aids to counteract hearing loss and the greater reliance on visual cues to compensate for declining speed of language processing in old age (Thompson, 1995).

Aside from the intended fit with the general lifespan scripts just described, two central motives were behind our proposal of SOC as a general psychological theory of behavior development: (a) to account for the realization of development in general and (b) to specify how individuals can effectively manage the overall lifespan changes in biological, psychological, and social conditions that form opportunities and constraints on levels and trajectories of development. In the sense of biocultural co-constructivism, the biogenetic and cultural contexts provide constraints and affordances (including interindividual differences in such constraints and affordances), and it is within these constraints that SOC operates.

In general, SOC component processes are considered to be universal. SOC-related behaviors, however, have the potential for a high degree of individual "phenotypic" specificity (P. B. Baltes & Baltes, 1990; Freund & Baltes, 2002). When expressed in the phenotypic sense, they show intra- and interindividual variability. Therefore, plasticity and its variable expression as a function of biocultural constraints is a cornerstone of SOC theory (P. B. Baltes & Singer, 2001; Lerner, 2002; S.-C. Li, 2003). Moreover, considering the triangulation of aspects of growth, maintenance, and loss, SOC can be viewed as an effective way to allocate and reallocate resources among these three functions.

In principle, SOC theory can be incorporated into many different theoretical perspectives, including behavioral-learning, biobehavioral, cognitive, action–theoretical (one of our preferred schemes), and social psychology (M. M. Baltes & Carstensen, 1996; Freund & Baltes, 2002; Marsiske, Lang, Baltes, & Baltes, 1995). Furthermore—and this reflects the many levels of consciousness and automaticity as well as external constraints that human behavior entails—SOC processes can vary along the dimensions active–passive, conscious–nonconscious, and internal–external. Along such lines, the SOC model can be applied to a variety of domains of functioning (e.g., social, cognitive, physical) and to different levels of analysis. For instance, the focus can be on a specific behavioral domain (e.g., working memory) or on personal functioning in a more general sense (e.g., subjective well-being or lifestyle). The focus can also entail how an institution, such as a school or nursing home, allocates its resources and staff behaviors to target aspects of growth, maintenance, or regulation of loss (M. M. Baltes, 1996).

A recent study conducted by Gignac, Cott, and Badley (2002) demonstrates the simultaneous occurrence of SOC as a general process and SOC as an individualized strategy of life management. Observational methods were used to study older patients afflicted with osteoarthritis and their strategies of management. The results showed that most participants made at least one adaptation that reflected either selection (e.g., restrict activity), optimization (e.g., practice movement), or compensation (e.g., use assistive devices). The fact that virtually all study participants did so reflects the universal aspect of SOC. Gignac et al. also reported large interindividual variability in the specific SOC behaviors expressed. This finding underscores the many variations that individuals can pursue as they produce their special ways of identifying and orchestrating ways of selecting, optimizing, and compensating.

Select Findings: Age Differences in SOC and Outcomes

Theory-guided research addressing questions about the application of SOC is just beginning. SOC-related behaviors can be assessed using self-report and observation methods and in experimental studies involving, for instance, the methodology of dual or multiple tasks. In this section, we summarize first findings from studies using a range of different methods. The outcomes carry a promissory spirit. First, there is evidence that the rank order and self-reported use of the SOC components change with age; second, there is evidence that people who engage in SOC behaviors show more adaptive outcomes.

As a developmental construct, we expect SOC to be an evolving system so that the behavioral repertoire associated with SOC reaches a peak somewhere in midlife. On the level of self-report, and as shown in Figure 2, initial findings support such a developmental gradient (Freund & Baltes, 2002). Young, middle-aged, and older adults answered a self-report instrument developed to assess preferred use of SOC strategies. Middle-aged adults reported the highest frequency of using all SOC components. In earlier and later phases of life, the SOC system seems less

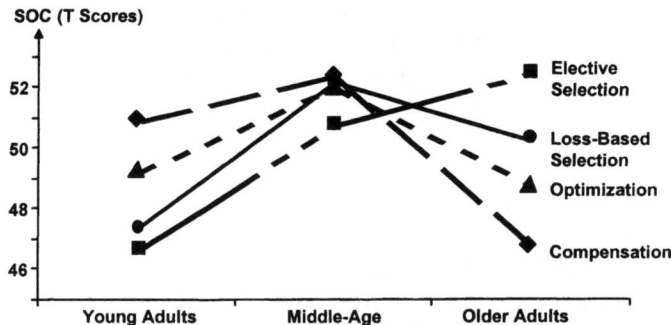

FIGURE 2   Age-group mean differences in four components of selective optimization with compensation (SOC; elective selection, loss-based selection, optimization, compensation): middle-aged adults reported the highest and perhaps most integrated endorsement of SOC (adapted from Freund & Baltes, 2002).

fully activated and coordinated. However, the age-related profiles carry meaning. For instance, the finding that older adults reported frequent use of elective selection corresponds to the view that aging individuals have fewer resources available and orient themselves increasingly toward fewer select goals (e.g., M. M. Baltes & Carstensen, 1996; P. B. Baltes & Baltes, 1990). Similarly, in young adulthood the task of life planning in a focused and concerted manner needs practice and refinement (e.g., Smith, 1999). Desires and volitions are less orchestrated.

The evidence available so far also suggests that the reported and observed use of SOC is associated with positive developmental outcomes. Where examined, the pattern of outcome correlations is robust against controlling for a number of rival predictors of positive development such as personality (e.g., the Big Five) and motivational constructs (e.g., tenacious goal pursuit and flexible goal adjustment). In samples ranging in age from 14 to 100+ years, it was found that adults who reported engaging in selection, optimization, and compensation when pursuing personal goals also reported higher levels of well-being (e.g., frequency of experiencing positive emotions, having a purpose in life, life satisfaction; Freund & Baltes, 1998, 2002; Wiese, Freund, & Baltes, 2000, 2002). In young adults, the evidence also includes reported success in dual-career partnerships and vocational advances (B. B. Baltes & Heydens-Gahir, 2003) as well as study behavior in college students (Wiese & Schmitz, 2002). In addition, Bajor and Baltes (2003) found that effective supervisors in work settings obtain better job performance ratings if they are seen to use strategies of SOC.

SOC and Dual-Task Research

SOC is a systemic theory. To deal with issues of biocultural co-construction and developmental contextualism, it is supposed to offer a window on coping with multiple contexts and multiple behavior demands. Rarely does human development involve a single task domain and a single context. Concurrent performances, such as being good at school and at sports, are more difficult than engaging in each of the tasks separately. SOC theory suggests that developmental researchers may want to use experimental paradigms developed for the study of dual- or multitask performance to better understand the developmental dynamics that individuals face as they regulate themselves in a complex time and context environment (Freund & Baltes, 2002; Krampe & Baltes, 2003; Lerner, Freund, de Stefanis, & Habermas, 2001; Lindenberger, Marsiske, & Baltes, 2000; see also various chapters in Staudinger & Lindenberger, 2003).

Experimental research on age differences in performance in dual tasks provides one concrete instantiation of predictions from SOC theory. For instance, when the dual task is to memorize a word list while walking fast or maintaining balance on a moving platform, the expectation is that, compared with young adults, older adults are more likely to prioritize walking or balance, because falling would be a more serious problem than not remembering a word in a list. This expectation that older adults show greater dual-task costs has been supported by findings from dual-task studies involving memorizing and walking carried out by K. Z. H. Li, Lindenberger, Freund, and Baltes (2001) and Lindenberger et al. (2000). Furthermore, older adults were effective in using compensatory skills to maintain a higher level of performance. Rapp, Krampe, and Baltes (2003) and Bondar, Krampe, and Baltes (2003) have reported similar findings from studies in which cognitive information processing and motor balance were competing tasks. These aging-associated effects of the differential use of SOC in favor of motor over cognitive performance are stronger when the behavioral system is tested at its limits (i.e., when the tasks were made more and more difficult).

SOC theory predicts that SOC behaviors have trait- and statelike characteristics. That this is so was shown in a study conducted by Bondar et al. (2003) involving motor and cognitive behavior. When it came to motor behavior and its high risk value, older adults showed preferential SOC behavior that was rather rigid. They did not reallocate resources when asked to do so. Regarding cognitive behavior, however, this was more easily possible. In light of the significance of maintaining motor function and balance, this asymmetrical allocation seems adaptive despite its apparent rigidity.

Differential allocation of resources can take many forms. Considering a different combination of tasks, namely, talking while walking, Kemper, Herman, and Lian (2003) demonstrated that older and younger adults differ in their compensatory strategies when task demands exceed their resources. Whereas young adults reduced the length and grammatical complexity of their spoken sentences, older adults reduced the rate of speech when they simultaneously had to walk. By speaking more slowly, older adults were able to preserve their speaking even under taxing dual-task conditions.

Taken together, these initial self-report and observational as well as experimental studies lend support to the perspective of the SOC theory of adaptive development. The replicated pattern of results suggests that individuals are better able to manage the tasks of life when they engage in selecting, optimizing, and compensating. Thus, SOC functions like a development-enhancing and loss-preventing general-purpose mechanism. As a general theory of adaptive development, it characterizes a system of strategies that permits individuals to master the general tasks of life, including those that result from the general lifespan scripts outlined earlier.

CONCLUSIONS

In this article, we have outlined a general frame for constructing developmental theory that is consistent with an overall biocultural architecture of human development. Our intent is to provide a frame that organizes the field as a whole. This organization proceeds from the general to the more specific across several levels of analysis. Our hope is that further explication of microgenetic and domain-specific processes are possible within this overarching meta-framework.

In this section, we mention one more theoretical conundrum that awaits new insights. It deals with the question whether development and aging can be viewed as part of the same framework or whether it is useful to treat these concepts as different entities. In general, our preference is to treat them from the same vantage point, or at least assume that they operate in conjunction—for instance, as an ongoing dynamic between gains and losses (P. B. Baltes, 1987). However, the recent explication of the biocultural architecture of ontogeny (P. B. Baltes, 1997) with its associated lifespan scripts highlights the possibility that there is much discontinuity between the causes and mechanisms of behavioral development at different stages of the lifespan. Given this, together with accumulating evidence about major losses of functioning in the Fourth Age, the oldest-old (P. B. Baltes & Smith, 1999, 2003), some might well ask whether the application of lifespan proposals about developmental processes to the end of life remains a tenable position.

Since 1990, much research has accumulated that addresses the potential and limits of ontogeny in old age and at the end of life. At present, two viewpoints pervade with regard to the interpretation of findings (e.g., P. B. Baltes, 1997; P. B. Baltes & Smith, 2003; Helmuth, 2003; Lachman, 2001). One is characterized by a spirit of scientific and social-policy optimism. Researchers who adopt this positive viewpoint highlight the history-graded advances in average life expectancy in developed countries together with the increasing opportunities for the majority of individuals in those societies to age successfully. The alternative standpoint tempers this optimism with reference to emergent uncertainties and challenges at the end of life (the Fourth Age). In particular, the positive news about human aging is called into question by findings about the oldest-old that indicate that their levels of physical, cognitive, emotional, and social functioning are much lower than those observed in the young-old. Research from the Berlin Aging Study (BASE) has illustrated this (P. B.

Baltes & Mayer, 1999; see also Smith, Maas, et al., 2002). Despite their optimistic reports on the young-old, BASE researchers uncovered some of the dilemmas and dysfunctionality of very old age. Data on 90- and 100-year-olds clearly show many aging-related losses, especially if the overall profile of aging trajectories is considered (e.g., Isaacowitz & Smith, 2003; Singer, Lindenberger, & Baltes, 2003; Smith & Baltes, 1997, 1998; Smith, Borchelt, Maier, & Jopp, 2002). BASE findings about increased dysfunctions in the oldest-old are the more significant as they apply to small subgroups of "positively" selected survivors; that is, to people who represent those few who survived into very old age and remained able to participate in the study. Thus, if anything, the measures collected in studies like BASE underestimate the actual plight of the oldest-old.

These findings generate concern about prospects for quality of life in very old age. They suggest that the chronic life strains experienced by the majority of the oldest-old gradually reduce the capacity of individuals to respond, adapt, and thrive. Stated otherwise—and in accord with theoretical proposals derived from the biocultural architecture of ontogeny and its associated lifespan scripts (P. B. Baltes, 1997)—there appears to be less continuity between the young-old and the Fourth Age than previously thought.

The lifespan psychological orientation to the study of aging evolved in part from numerous discussions during the 1950s and 1960s about the definition of *development* and the relation between development and aging (e.g., Anderson, 1958; Bayley, 1963; Birren, 1964). At that time, developmental change in early life (defined primarily in terms of growth, increasing organization, and structural differentiation) had been viewed by some as conceptually different from behavioral change associated with aging (described as decline, disorganization, dedifferentiation). Two key conferences, one arranged by researchers of child development (see Harris, 1957) and the other by researchers interested in adulthood and aging (see Birren, 1964), discussed methodological issues and examined prospects for theoretical and considerations of development across the lifespan. During the subsequent decades, the lifespan psychological orientation established a conceptual bridge that contributed to a broadening of the questions posed about the processes and contextual influences on development from conception to death (e.g., P. B. Baltes, 1987, 1997; Hetherington & Baltes, 1988). Perhaps, at the start of a new century, and especially with regard to the end of life (the Fourth Age), it might be worth re-examining the relations between concepts of development and concepts of aging.

ACKNOWLEDGMENTS

We gratefully acknowledge countless valuable discussions and contributions from all current and former colleagues in the Center for Lifespan Psychology, especially Alexandra M. Freund, Shu-Chen Li, Ulman Lindenberger, and Ursula M. Staudinger.

REFERENCES

Anderson, J. E. (1958). A developmental model for aging. *Vita Humana, 1,* 5–18.
Bäckman, L., Small, B. J., Wahlin, A., & Larsson, M. (2000). Cognitive functioning in very old age. In F. I. M. Craik & T. A. Salthouse (Eds.), *The handbook of aging and cognition* (2nd ed., pp. 499–558). Mahwah, NJ: Lawrence Erlbaum Associates, Inc.
Bajor, J. K., & Baltes, B. B. (2003). The relationship between selective optimization with compensation, conscientiousness, motivation, and performance. *Journal of Vocational Behavior, 63,* 347–367.
Baltes, B. B., & Heydens-Gahir, H. A. (2003). Reduction of work–family conflict through the use of selection, optimization, and compensation behaviors. *Journal of Applied Psychology, 88,* 1005–1018.
Baltes, M. M. (1996). *The many faces of dependency in old age.* Cambridge, England: Cambridge University Press.
Baltes, M. M., & Carstensen, L. L. (1996). The process of successful ageing. *Ageing and Society, 16,* 397–422.
Baltes, P. B. (1987). Theoretical propositions of lifespan developmental psychology: On the dynamics between growth and decline. *Developmental Psychology, 23,* 611–626.
Baltes, P. B. (1997). On the incomplete architecture of human ontogeny: Selection, optimization, and compensation as foundation of developmental theory. *American Psychologist, 52,* 366–380.
Baltes, P. B., & Baltes, M. M. (1990). Psychological perspectives on successful aging: The model of selective optimization with compensation. In P. B. Baltes & M. M. Baltes (Eds.), *Successful aging: Perspectives from the behavioral sciences* (pp. 1–34). Cambridge, England: Cambridge University Press.
Baltes, P. B., Cornelius, S. W., & Nesselroade, J. R. (1979). Cohort effects in developmental psychology. In J. R. Nesselroade & P. B. Baltes (Eds.), *Longitudinal research in the study of behavior and development* (pp. 61–87). New York: Academic.
Baltes, P. B., Freund, A. M., & Li, S.-C. (in press). The psychological science of human ageing. In M. Johnson (Ed.), *The Cambridge handbook of age and ageing.* Cambridge, England: Cambridge University Press.
Baltes, P. B., Lindenberger, U., & Staudinger, U. M. (1998). Lifespan theory in developmental psychology. In R. M. Lerner (Ed.), *Handbook of child psychology: Vol. 1. Theoretical models of human development* (5th ed., pp. 1029–1143). New York: Wiley.
Baltes, P. B., & Mayer, K. U. (Eds.). (1999). *The Berlin Aging Study: Aging from 70 to 100.* New York: Cambridge University Press.
Baltes, P. B., & Nesselroade, J. R. (1984). Paradigm lost and paradigm regained: Critique of Dannefer's portrayal of life-span developmental psychology. *American Sociological Review, 49,* 841–846.
Baltes, P. B., Reese, H. W., & Lipsitt, L. P. (1980). Lifespan developmental psychology. *Annual Review of Psychology, 31,* 65–110.
Baltes, P. B., Reese, H. W., & Nesselroade, J. R. (1977). *Lifespan developmental psychology: Introduction to research methods.* Hillsdale, NJ: Lawrence Erlbaum Associates, Inc.
Baltes, P. B., & Schaie, K. W. (1976). On the plasticity of intelligence in adulthood and old age: Where Horn and Donaldson fail. *American Psychologist, 31,* 720–725.

Baltes, P. B., & Singer, T. (2001). Plasticity and the ageing mind: An exemplar of the bio-cultural orchestration of brain and behaviour. *European Review, 9,* 59–76.
Baltes, P. B., & Smith, J. (1999). Multilevel and systemic analyses of old age: Theoretical and empirical evidence for a fourth age. In V. L. Bengtson & K. W. Schaie (Eds.), *Handbook of theories of aging* (pp. 153–173). New York: Springer.
Baltes, P. B., & Smith, J. (2003). New frontiers in the future of aging: From successful aging of the young old to dilemmas of the fourth age. *Gerontology, 49,* 123–135.
Bandura, A. (1982). The psychology of chance encounters and life paths. *American Psychologist, 37,* 747–755.
Bayley, N. (1963). The life span as a frame of reference in psychological research. *Vita Humana, 6,* 125–139.
Birren, J. E. (Ed.). (1964). *Relations of development and aging.* Springfield, IL: Charles C Thomas.
Bondar, A., Krampe, R. T., & Baltes, P. B. (2003). *Balance takes priority over cognition: Can young and older adults deliberately control resource allocation?* Berlin, Germany: Max Planck Institute for Human Development.
Brandtstädter, J. (1998). Action perspectives on human development. In R. M. Lerner (Ed.), *Handbook of child psychology: Vol. 1. Theoretical models of human development* (5th ed., pp. 807–866). New York: Wiley.
Brandtstädter, J., & Lerner, R. M. (Eds.). (1999). *Action and self-development: Theory and research through the life span.* Thousand Oaks, CA: Sage.
Brim, O. G., Jr., & Ryff, C. D. (1980). On the properties of life events. In P. B. Baltes & O. G. Brim, Jr. (Eds.), *Lifespan development and behavior* (Vol. 3, pp. 367–388). New York: Academic.
Cantor, N., & Fleeson, W. (1994). Social intelligence and intelligent goal pursuit: A cognitive slice of motivation. In W. D. Spaulding (Ed.), *Nebraska Symposium on Motivation: Integrative views of motivation, cognition, and emotion* (Vol. 41, pp. 125–179). Lincoln: University of Nebraska Press.
Carstensen, L. L. (1995). Evidence for a life-span theory of socioemotional selectivity. *Current Directions in Psychological Science, 4,* 151–156.
Dannefer, D. (1992). On the conceptualization of context in developmental discourse: Four meanings of context and their implications. In D. L. Featherman, R. M. Lerner, & M. Perlmutter (Eds.), *Lifespan development and behavior* (Vol. 11, pp. 83–110). Hillsdale, NJ: Lawrence Erlbaum Associates, Inc.
Dannefer, D., & Uhlenberg, P. (1999). Paths of the life course: A typology. In V. L. Bengtson & K. W. Schaie (Eds.), *Handbook of theories of aging* (pp. 306–326). New York: Springer.
Diehl, M. (1999). Self-development in adulthood and aging: The role of critical life events. In C. D. Ryff & V. W. Marshall (Eds.), *The self and society in aging processes* (pp. 150–183). New York: Springer.
Dixon, R. A., & Bäckman, L. (Eds.). (1995). *Compensating for psychological deficits and declines: Managing losses and promoting gains.* Mahwah, NJ: Lawrence Erlbaum Associates, Inc.
Durham, W. H. (1991). *Coevolution: Genes, culture and human diversity.* Stanford, CA: Stanford University Press.
Eccles, J. S., & Wigfield, A. (2002). Motivational beliefs, values, and goals. *Annual Review of Psychology, 53,* 109–132.
Elbert, T. (2003, August). *Significant life events and psychological trauma: Consequences for brain structure and function.* Paper presented at the Conference on Brain, Mind, and Culture: From Interactionism to Biocultural Co-Constructivism, Döllnsee, Germany.
Elder, G. H. (1998). The life-course and human development. In R. M. Lerner (Ed.), *Handbook of child psychology: Vol. 1. Theoretical models of human development* (5th ed., pp. 939–991). New York: Wiley.
Ferraro, K. F. (2001). Aging and role transitions. In R. H. Binstock & L. K. George (Eds.), *Handbook of aging and the social sciences* (5th ed., pp. 313–330). San Diego, CA: Academic.
Finch, C. E. (1990). *Longevity, senescence, and the genome.* Chicago: University of Chicago Press.

Finch, C. E., & Kirkwood, T. B. L. (2000). *Chance, development, and aging.* New York: Oxford University Press.
Flynn, J. R. (1999). Searching for justice: IQ gains over time. *American Psychologist, 54,* 5–20.
Freund, A. M., & Baltes, P. B. (1998). Selection, optimization, and compensation as strategies of life-management: Correlations with subjective indicators of successful aging. *Psychology and Aging, 13,* 531–543.
Freund, A. M., & Baltes, P. B. (2002). Life-management strategies of selection, optimization, and compensation: Measurement by self-report and construct validity. *Journal of Personality and Social Psychology, 82,* 642–662.
Gignac, M. A. M., Cott, C., & Badley, E. M. (2002). Adaptation to disability: Applying selective optimization with compensation to the behaviors of older adults with osteoarthritis. *Psychology and Aging, 17,* 520–524.
Gollin, E. S. (1981). Development and plasticity. In E. S. Gollin (Ed.), *Developmental plasticity: Behavioral and biological aspects of variations in development* (pp. 231–251). New York: Academic.
Goulet, L. R., & Baltes, P. B. (Eds.). (1970). *Lifespan developmental psychology: Research and theory.* New York: Academic.
Harris, D. E. (1957). *The concept of development: An issue in the study of human behavior.* Minneapolis: University of Minnesota Press.
Havighurst, R. J. (1972). *Developmental tasks and education* (3rd ed.). New York: McKay.
Heckhausen, J., Dixon, R. A., & Baltes, P. B. (1989). Gains and losses in development throughout adulthood as perceived by different adult age groups. *Developmental Psychology, 25,* 109–121.
Heckhausen, J., & Dweck, C. S. (Eds.). (1998). *Motivation and self-regulation across the life span.* New York: Cambridge University Press.
Heckhausen, J., & Schulz, R. (1995). A life-span theory of control. *Psychological Review, 102,* 284–304.
Helmuth, L. (2003, February 28). The wisdom of the wizened. *Science, 299,* 1300–1302.
Hetherington, E. M., & Baltes, P. B. (1988). Child psychology and lifespan development. In E. M. Hetherington, R. M. Lerner, & M. Perlmutter (Eds.), *Child development in lifespan perspective* (pp. 1–19). Hillsdale, NJ: Lawrence Erlbaum Associates, Inc.
Isaacowitz, D. M., & Smith, J. (2003). Positive and negative affect in very old age. *Journal of Gerontology: Psychological Sciences, 58B,* P143–P152.
Kemper, S., Herman, R. E., & Lian, C. H. T. (2003). The costs of doing two things at once for young and older adults: Talking while walking, finger tapping, and ignoring speech or noise. *Psychology and Aging, 18,* 181–192.
Kirkwood, T. B. L. (2002). Evolution of ageing. *Mechanisms of Ageing and Development, 123,* 737–745.
Kirkwood, T. B. L. (2003). Age differences in evolutionary selectional benefits. In U. M. Staudinger & U. Lindenberger (Eds.), *Understanding human development: Dialogues with lifespan psychology* (pp. 215–241). Dordrecht, The Netherlands: Kluwer.
Kliegl, R., Smith, J., & Baltes, P. B. (1989). Testing-the-limits and the study of adult age differences in cognitive plasticity of a mnemonic skill. *Developmental Psychology, 25,* 247–256.
Krampe, R. T., & Baltes, P. B. (2003). Intelligence as adaptive resource development and resource allocation: A new look through the lenses of SOC and expertise. In R. J. Sternberg & E. L. Grigorenko (Eds.), *Perspectives on the psychology of abilities, competencies, and expertise* (pp. 31–69). New York: Cambridge University Press.
Kruse, A., Lindenberger, U., & Baltes, P. B. (1993). Longitudinal research on human aging: The power of combining real-time, microgenetic, and simulation approaches. In D. Magnusson & P. Casaer (Eds.), *Longitudinal research on individual development: Present status and future perspectives* (pp. 153–193). Cambridge, England: Cambridge University Press.
Lachman, M. E. (2001). *Handbook of midlife development.* New York: Wiley.

Lerner, R. M. (1984). *On the nature of human plasticity*. Cambridge, England: Cambridge University Press.
Lerner, R. M. (2002). *Concepts and theories of human development*. Mahwah, NJ: Lawrence Erlbaum Associates, Inc.
Lerner, R. M., Freund, A. M., de Stefanis, I., & Habermas, T. (2001). The selection, optimization, and compensation model as a frame for understanding developmental regulation in adolescence. *Human Development, 44*, 29–50.
Li, K. Z. H., Lindenberger, U., Freund, A. M., & Baltes, P. B. (2001). Walking while memorizing: Age-related differences in compensatory behavior. *Psychological Science, 12*, 230–237.
Li, S.-C. (2003). Biocultural orchestration of developmental plasticity across levels: The interplay of biology and culture in shaping the mind and behavior across the lifespan. *Psychological Bulletin, 129*, 171–194.
Lindenberger, U., & Baltes, P. B. (1995). Testing-the-limits and experimental simulation: Two methods to explicate the role of learning in development. *Human Development, 38*, 349–360.
Lindenberger, U., & Baltes, P. B. (2000). Lifespan psychology theory. In A. E. Kazdin (Ed.), *Encyclopedia of psychology* (Vol. 5, pp. 52–57). New York: Oxford University Press.
Lindenberger, U., Marsiske, M., & Baltes, P. B. (2000). Memorizing while walking: Increase in dual-task costs from young adulthood to old age. *Psychology and Aging, 15*, 417–436.
Magnusson, D. (Ed.). (1996). *The lifespan development of individuals: Behavioural, neurobiological, and psychosocial perspectives*. Cambridge, England: Cambridge University Press.
Magnusson, D., & Casaer, P. J. M. (Eds.). (1993). *Longitudinal research on individual development: Present status and future perspectives*. Cambridge, England: Cambridge University Press.
Maier, H., & Vaupel, J. (2003). Age differences in cultural efficacy: Secular trends in longevity. In U. M. Staudinger & U. Lindenberger (Eds.), *Understanding human development: Dialogues with lifespan psychology* (pp. 59–78). Boston: Kluwer Academic.
Manton, K. G., Stallard, E., & Corder, L. (1997). Changes in the age dependence of mortality and disability: Cohort and other determinants. *Demography, 34*, 135–157.
Marsiske, M., Lang, F. R., Baltes, P. B., & Baltes, M. M. (1995). Selective optimization with compensation: Life-span perspectives on successful human development. In R. A. Dixon & L. Bäckman (Eds.), *Psychological compensation: Managing losses and promoting gains* (pp. 35–79). Mahwah, NJ: Lawrence Erlbaum Associates, Inc.
Mayer, K. U. (2003). The sociology of the life course and lifespan psychology: Diverging or converging pathways? In U. M. Staudinger & U. Lindenberger (Eds.), *Understanding human development: Dialogues with lifespan psychology* (pp. 463–481). Dordrecht, The Netherlands: Kluwer.
McAdams, D. P. (2001). The psychology of life stories. *Review of General Psychology, 5*, 100–122.
McClearn, G. E. (1997). Biomarkers of age and aging. *Experimental Gerontology, 32*, 87–94.
Modell, J., & Elder, G. H., Jr. (2002). Children develop in history: So what's new? *Minnesota Symposium on Child Psychology, 32*, 173–205.
Neugarten, B. L. (1969). Continuities and discontinuities of psychological issues into adult life. *Human Development, 12*, 121–130.
Rapp, M., Krampe, R. T., & Baltes, P. B. (2003). *Preservation of skills in Alzheimer's disease: The case of postural control*. Berlin, Germany: Max Planck Institute for Human Development.
Riley, M. W., Foner, A., & Riley, J. W. (1999). The aging and society paradigm. In V. L. Bengtson & K. W. Schaie (Eds.), *Handbook of theories of aging* (pp. 327–343). New York: Springer.
Rosow, I. (1985). Status and role change through the life cycle. In R. H. Binstock & E. Shanas (Eds.), *Handbook of aging and the social sciences* (2nd ed., pp. 62–93). New York: Van Nostrand Reinhold.
Sampson, R. J., Morenoff, J. D., & Gannon-Rowley, T. (2002). Assessing "neighborhood effects": Social processes and new directions in research. *Annual Review of Sociology, 28*, 443–478.

Schaie, K. W. (1965). A general model for the study of developmental problems. *Psychological Bulletin, 64*, 92–107.
Schaie, K. W. (1996). *Intellectual development in adulthood: The Seattle Longitudinal Study*. New York: Cambridge University Press.
Settersten, R. A. (1999). *Lives in time and place: The problems and promises of developmental science*. New York: Baywood.
Shanahan, M. J. (2000). Pathways to adulthood in changing societies: Variability and mechanisms in life course perspective. *Annual Review of Sociology, 26*, 667–692.
Singer, T., Lindenberger, U., & Baltes, P. B. (2003). Plasticity of memory for new learning in very old age: A story of major loss. *Psychology and Aging, 18*, 306–317.
Smith, J. (1999). Life planning: Anticipating future life goals and managing personal development. In J. Brandtstädter & R. M. Lerner (Eds.), *Action and self-development: Theory and research through the life span* (pp. 223–255). Thousand Oaks, CA: Sage.
Smith, J. (2001). Old age and centenarians. In N. Smelser & P. B. Baltes (Eds.), *International encyclopedia of the social and behavioral sciences* (pp. 10843–10847). Oxford, England: Elsevier.
Smith, J. (2003). The gain–loss dynamic in lifespan development: Implications for change in self and personality during old and very old age. In U. M. Staudinger & U. Lindenberger (Eds.), *Understanding human development: Dialogues with lifespan psychology* (pp. 215–241). Dordrecht, The Netherlands: Kluwer.
Smith, J., & Baltes, P. B. (1997). Profiles of psychological functioning in the old and oldest old. *Psychology and Aging, 12*, 458–472.
Smith, J., & Baltes, M. M. (1998). The role of gender in very old age: Profiles of functioning and everyday life patterns. *Psychology and Aging, 13*, 676–695.
Smith, J., Borchelt, M., Maier, H., & Jopp, D. (2002). Health and well-being in old age. *Journal of Social Issues, 58*, 715–732.
Smith, J., Maas, I., Mayer, K. U., Helmchen, H., Steinhagen-Thiessen, E., & Baltes, P. B. (2002). Two-wave longitudinal findings from the Berlin Aging Study: Introduction to a collection of papers. *Journal of Gerontology: Psychological Sciences, 57B*, P471–P473.
Staudinger, U. M., & Lindenberger, U. (Eds.). (2003). *Understanding human development: Dialogues with lifespan psychology*. Dordrecht, The Netherlands: Kluwer.
Staudinger, U. M., Marsiske, M., & Baltes, P. B. (1993). Resilience and levels of reserve capacity in later adulthood: Perspectives from life-span theory. *Development and Psychopathology, 5*, 541–566.
Tetens, J. N. (1777). *Philosophische Versuche über die menschliche Natur und ihre Entwicklung* [Philosophical essays on human nature and its development]. Leipzig, Germany: Weidmanns Erben und Reich.
Thompson, L. A. (1995). Encoding and memory for visible speech and gestures: A comparison between young and older adults. *Psychology and Aging, 10*, 215–228.
Uhlenberg, P. (1988). Aging and the societal significance of cohorts. In J. E. Birren & V. L. Bengtson (Eds.), *Emergent theories of aging* (pp. 405–425). New York: Springer.
Uttal, D. H., & Perlmutter, M. (1989). Toward a broader conceptualization of development: The role of gains and losses across the life span. *Developmental Review, 9*, 101–132.
Vaupel, J. W., Carey, J. R., Christensen, K., Johnson, T. E., Yashin, A. I., Holm, N. V., et al. (1998, May 8). Biodemographic trajectories of longevity. *Science, 280*, 855–860.
Waddington, C. H. (1966). *Principles of development and differentiation*. New York: Macmillan.
Wiese, B. S., Freund, A. M., & Baltes, P. B. (2000). Selection, optimization, and compensation: An action-related approach to work and partnership. *Journal of Vocational Behavior, 57*, 273–300.
Wiese, B. S., Freund, A. M., & Baltes, P. B. (2002). Subjective career success and emotional well-being: Longitudinal predictive power of selection, optimization and compensation. *Journal of Vocational Behavior, 60*, 321–335.

Wiese, B. S., & Schmitz, B. (2002). Studienbezogenes Handeln im Kontext eines entwicklungspsychologischen Meta-Modells [Action regulation at university: Application of a developmental meta-model]. *Zeitschrift für Entwicklungspsychologie und Pädagogische Psychologie, 34,* 80–94.

Wrosch, C., & Freund, A. M. (2001). Self-regulation of normative and nonnormative developmental challenges. *Human Development, 44,* 264–283.

The Interpenetration of Culture and Biology in Human Development

Cynthia García Coll
Brown University

The central argument of this article is that human development should be viewed as the product of the interpenetration[1] of cultural and biological processes. Using developmental system theory, it is argued that our work as investigators of human development is to examine this mutual coregulation: how biology and culture interpenetrate over time in concert with developmental processes. Various findings from developmental research are used to provide evidence on (a) how our biology constrains, gives expression to, mediates, or moderates how culture operates on developmental processes and (b) how culture infiltrates and becomes part of our biology and basic developmental processes. Examples of advances in neuroscience are also used to illustrate this interpenetration and to argue for the pervasiveness of cultural influences on development. To conduct the necessary research, new methodologies are needed. Promise is seen in emerging areas of social cognitive neuroscience.

There's an elephant in the room. It is large and squatting, so it is hard to get around it. Yet we squeeze by with, "How are you?" and "I'm fine," and a thousand other forms of trivial chatter. We talk about the weather. We talk about work. We talk about everything else, except the elephant in the room.
—*The Elephant in the Room*, author unknown

The importance of culture on life span development is like the elephant in the room. At a certain level, we all know that culture is important, but most of us disregard it in our day-to-day theorizing and studying of developmental processes. Those of us who do not want to see the elephant—or, for that matter, acknowledge it at all—go to the other extreme and proclaim that evolution and biology are the primordial, if

---

Requests for reprints should be sent to Cynthia García Coll, Brown University, Box 1938, Providence, RI 02912. E-mail: Cynthia_Garcia_Coll@brown.edu

[1]I have borrowed this term from Lewontin (1982), as cited by Oyama (2000).

not the only, sources of important influences in developmental processes. Others who want to make us aware that the elephant really does exist engage in painstaking advocacy to convince the rest of us that culture really matters and that biology, for the most part, is irrelevant. Thus the pendulum of the nature–nurture controversy keeps moving in its own rhythmic reiteration in space and time.

Since the beginning of my own developmental career, I have occupied that peculiar space where both culture[2] and biology[3] matter and get intertwined and expressed in most developmental processes. In my first published article, "Cultural and Biomedical Correlates of Neonatal Behavior" (García Coll, Sepkoski, & Lester, 1981), I argued that both realms of influences and their interactions could be identified as early as the first day of life. Forced by the need to isolate, to bifurcate, to measure "independent or relative contributions" that is so prevalent in the field, I have found myself repeatedly going over to one end of the continuum versus the other, isolating culture or biology in everything I studied. Yet my theoretical positions (i.e., García Coll, 1990; García Coll et al., 1996) have always incorporated both at some point in my analyses (see also García Coll, Bearer, & Lerner, 2004). Not comfortable in either camp, I have straddled the line in between, always feeling neither here nor there.

Now influenced by recent empirical and theoretical work on developmental, dynamic systems approaches,[4] I have found a more comfortable place that captures my evolving view of how culture operates on developmental processes. My purpose in this article is to unravel a view that is very much still in progress: a developmental, dynamic system approach to the consideration of culture on human development. I use the word *biocultures* to denote that system. This article is not intended to provide an overview of research in any particular area;[5] rather, my intention is to use examples from many different areas of research to illustrate a series of theoretical points that I hope will lead to some new directions of thought and inquiry in this important area.

---

[2]By *culture* I mean both the more traditional definition of symbols, values, and meaning-making systems and traditions of particular societies that share historical roots as well as the more "modern" one that emphasizes the daily routines and practices that enact such systems.

[3]By *biology* I mean any embodied system, from gene expression and other biochemical processes to the biological embedding that results from the impact primarily of early experience on developing biological systems.

[4]I am indebted to Anne Fausto-Sterling and other colleagues in the Pembroke Center's Seminar on Theories of Embodiment (2002–2003) held at Brown University, the theoretical work of Susan Oyama, and the work by Richard Lerner and Esther Thelen on human development and Gilbert Gottlieb in comparative psychology.

[5]In the process of writing this article I came across two excellent expositions of similar theoretical orientations to the study of culture and biology on human development that provide excellent reviews of supportive literature across the life span (Kennepohl, 1999; Li, 2003).

WHAT IS A DEVELOPMENTAL, DYNAMIC SYSTEMS APPROACH?

(Human) Development can only be understood as the multiple, mutual, and continuous interaction of all the levels of the developing system, from the molecular to the cultural. (Thelen & Smith, 1998, p. 563)

Recent writings by Thelen and Smith (1998), Lerner (2004), Gottlieb (2004), Oyama (2000), and Fausto-Sterling (1992, 2000, 2003) provide excellent overviews of the intellectual and empirical origins of developmental system theories as well as examples of the variety of current applications. For my purpose, I use some principles from the theoretical positions that I find useful in understanding how culture operates on life span development.

A dynamic systems approach to human development emphasizes the continuing "transactions within and between cells, within and between components of the body and environment ... in the regulation of developmental change" (Fogel, 1993, p. 49). This conceptualization has profound implications for how researchers should study developmental processes throughout the life span and the multiple, interactive, dynamic influences that accompany that journey. It implies that even if one is studying a single source of influence, one would acknowledge at some critical point that that source is not acting in isolation.

A graphical illustration by Gottlieb (1992, p. 241) simplifies such complexity into four levels of analyses: (a) genetic, (b) neural, (c) behavioral (which also encompasses psychological functioning; see Gottlieb, Wahlsten, & Lickliter, 1998), and (d) environmental (physical, social, and cultural). This schematic implies that the isolation of one level of analyses for the heuristic purpose of scientific investigation does not entail its isolation in function. Taking this schematic seriously implies that just as one cannot presume that biology is not part of the developmental trajectory over the life span, so too can one not assume that one's culture is not. Even if one isolates and studies single influences for heuristic purposes, finding significant effects of one does not preclude the influence of others.

However, a dynamic systems approach not only connotes a coexistence of influences but also goes beyond having independent, main effects or even interactions in statistical terms. It connotes a close, intertwined, codetermined, coregulation and therefore a mutually created system among all levels (Overton, 1998). Figure 2 on page 125 of Overton (1998) provides a schematic of what Overton called a *self-organizing system*: the bio/social–cultural action matrix, in which the biological as well as the sociocultural realms are constantly interpenetrating and interdependently emerging subsystems. In this conceptualization not only do biological processes depend on the environment, but culture also depends on biological processes. A child cannot learn a language well without a neural substrate that processes the input—or, equally important, without an environment that provides the input. Each new word is a function of both; perhaps sometimes the influence of one versus the other is more evident, but it never works in isolation.

Our work as investigators of human development is to examine this mutual coregulation: how biology and culture interpenetrate over time in concert with developmental processes. Neither one influence singly determines them; both matter (Rogoff, 2003). Although all of the previously cited investigators and theoreticians acknowledge this basic interplay, to my knowledge we all have failed to study simultaneously both biological and cultural processes and their interpenetration. My main point is that as long as we continue to isolate, to bifurcate these realms, our understanding of human developmental processes will be limited.

THE BIOLOGY IN CULTURE: HOW OUR BIOLOGY CONSTRAINS, GIVES EXPRESSION TO, MEDIATES, OR MODERATES HOW CULTURE OPERATES ON DEVELOPMENTAL PROCESSES

There are at least two ways that our biology impinges on the way culture operates on human development. One way is how our bodies "experience" environmental inputs. What we see, hear, or experience from our cultural milieus is in part a function of our perceptual apparatus and our capacities to encode, transform, and build on environmental input. Constraints are seen, for example, in our limited night vision, which has led to inventions such as sources of light and the organization of appropriate day and night activities. Possibilities of expression are seen, for example, in the ability of our vocal cords to create the sounds associated with language.

However, as our theoretical stance predicts, in turn, the functional capacity of our perceptual and information-processing apparatus is in part a reflection of environmental inputs, inclusive of culture. This continuous interdependence clearly exemplifies the dynamical aspects of human development. A good example of this interplay comes from research on speech perception. Although speech perception begins early in life as a function in part of a perceptual system that has relative sensitivities and insensitivities (e.g., perception of categorical phonemes; see Spelke & Newport, 1998), changes in language perception abilities are seen very early in life as a function of linguistic input. Work by Werker and her colleagues (Werker & Lalonde, 1988; Werker & Tees, 1984) has shown how the abilities to differentiate the phonemes that are universally present at birth are lost between ages 6 to 12 months as a function of the contrastive categories that are present in the surrounding linguistic systems. This loss is observed in adults, unless training is instituted to reverse it. Because language is considered one of the most powerful transmissions of culture, and linguistic environments are so culturally bound (Schieffelin, 1990; Schieffelin & Ochs, 1986; Shweder et al., 1998), this loss of perceptual discriminatory abilities is one of the earliest evidence of how culture becomes embedded into our developing perceptual systems.

The other way that biology impinges on cultural influences is by creating behavioral predispositions that are conceptualized as resulting from evolutionary processes that reflect thousands of years of organismic response to environmental pressures. For example, in evolutionary terms, we are social animals: We are attracted to socially relevant stimuli (i.e., human faces and voices), we get attached to significant others, we communicate with others, and we live in groups (Cynader & Frost, 1999). These biological predispositions to a certain extent predispose us to acquire cultural tools (Grossmann, 1995; Rogoff, 2003). For example, the development of an attachment to a primary caregiver figure is one of the most important evolutionary predispositions that we share with many other species (Bowlby, 1969). Indeed, infants all over the world, in cultures as distant as kibbutzim, African bush, Japan, and urban Germany, demonstrate attachments to primary caregiver figures by the end of the 1st year of life (van Ijzendoorn & Kroonenberg, 1988). As Grossmann (1995) asserted, "Viewed properly, attachment is the very foundation for a child's ability to understand and participate in the extended social and cultural world" (pp. 92–93). Our biological predispositions enable us to become cultural beings. However, as I discuss in the next section, even this "universal" process is expressed and moderated by the specificity of culturally defined contexts.

THE CULTURE IN BIOLOGY: HOW DOES CULTURE INFILTRATE AND BECOME PART OF OUR BIOLOGY AND DEVELOPMENTAL PROCESSES?

Recent conceptualizations of cultural influences on human development emphasize the importance of activity settings and of daily routines in the enculturation process of children (i.e., Rogoff, 2003; Weisner, 2002). As Weisner (2002) asserted, "(they) crystallize culture directly in everyday experience, because they include values and goals, resources needed to make the activity happen, people in relationships, the tasks the activity is there to accomplish, emotions and motives of those engaged in the activity, and a script defining the appropriate, normative way to engage in that activity" (p. 275). Following this theoretical stance, these daily activities are conceptualized as one of the mechanisms by which culture becomes embedded in our developmental processes and build on existing biological predispositions.

A good example is provided again by the research on attachment. One of the most basic influences on the development of attachment during the 1st year of life is the pattern of daily mother–infant interactions (Ainsworth, 1967; Ainsworth, Blehar, Waters, & Wall, 1978). Many studies have documented the relation between patterns of daily mother–infant interaction and attachment classification in many different cultures around the world (Ainsworth et al., 1978; Bretherton, 1995). At the same time, patterns of mother–infant interaction are one of the most basic transmissions of cultural values (Bornstein & Cote, 2001; Bornstein, Azuma, Tamis-LeMonda, & Ogino, 1990; Rogoff, 2003). For example, Harwood, Schoelmerich, Schulze, and Gonzalez (1999) found that mother–infant interaction patterns differed among Puerto Rican and Anglo mothers, whereby the Puerto Rican mothers pose fewer questions and are more directive in their interactions than the Anglo mothers. These patterns of interactions reflect the high value placed on *respeto* and strict age roles in Puerto Rican culture as opposed to the Anglo-American culture.

Perhaps because of these culturally guided social interactions, we see differences in the distribution of classifications of infants' attachment around the world. Cross-cultural differences are seen in the distribution of attachment classifications (Grossmann, Grossmann, Spangler, Suess, & Unzer, 1985; Miyake, Chen, & Campos, 1985; Sagi et al., 1985). Although the authors' interpretation of these differences as a reflection of cross-cultural differences in the distribution of attachment classifications has been questioned (Bretherton, 1995), the data on the variety of caregiving environments for infants around the world suggest that different cultural practices and parental beliefs might moderate the development of attachment classifications. The most commonly observed differences are in providing additional attachment figures or creating other ways of developing a secure attachment than exclusive, close, sensitive, consistent interactions with one's biological mother (Tronick, Winn, & Morelli, 1985). These observations led Bretherton (1995) to assert: "In sum, attachment behavior is heavily overlaid with cultural prescriptions, even in a society that much more closely resembles the conditions of human evolution than our own" (p. 75).

Another way that cultures operate on biological processes is by providing meaning for developmental processes that vary across cultures. For example, so-called universal infant attachment behaviors are interpreted differently as a function of cultural values. Harwood (1992) compared lower class Puerto Rican mothers' and lower and middle-class Anglo mothers' description of desirable and undesirable attachment behaviors. Anglo mothers, irrespective of social class, placed significantly greater value than did Puerto Rican mothers on behavioral characteristics of the children that reflected the development of personal abilities (i.e., secure). In contrast, Puerto Rican mothers emphasized behaviors associated with the maintenance of proper respect and demeanor (i.e., calm, obedient, and well brought up), which are highly valued in both children and adults in that culture. Again, a close interplay between biological and cultural processes is observed.

WHAT DO ADVANCES IN NEUROSCIENCE TELL US ABOUT THE INTERPENETRATION OF BIOLOGY AND ENVIRONMENT AND THE POTENTIAL FOR CULTURAL INFLUENCES ON DEVELOPMENT?

How do these biological–cultural developmental processes become embodied in our biologically originated structures, functions, and processes? The term *biological embedding* has been used by some investigators to refer to how experiences become part of our biology. Keating and Hertzman (1999) defined *biological embedding* as a process "whereby systematic differences in psychosocial/material circumstances, from conception onwards, embed themselves in human biology" (p. 11). But how does that embedding take place?

Advances in neuroscience have brought about a clearer understanding of how brain structures are shaped by environmental input (see Fausto-Sterling, 2000). Phrases such as *neuronal sculpting, neuronal death*, and *synaptic pruning* are currently used to describe these epigenetic processes (Cynader & Frost, 1999). Cynader and Frost (1999, p. 154) asserted that "during development, information from genetic sources, the material environment, and biological and social environments *all* [italics added]" contribute in complementary ways and at critical times during neural differentiation *to forge competencies for the current ecology of the individual* [italics added]. In this framework, I argue that the ecologically meaningful competencies are defined in part by cultural prescriptions embedded in daily routines and meanings.

There are two ways that the brain is affected by early experience: (a) structurally and (b) neurochemically. Greenough and Black (Black & Greenough, 1986; Greenough & Black, 1992) coined the phrases *experience expectant* and *experience dependent* to describe the two ways that interactions with the environment guide brain development. There is extensive evidence with animal models on how certain somatosensory parts of the cortex are dependent on particular environmental input for it to proceed along normative developmental pathways (Cynader & Frost, 1999; Gottlieb, Wahlsten, & Lickliter, 1998). In particular, the visual cortex is highly dependent on exposure of light and patterned visual information during a sensitive period for its subsequent normative development. When certain amounts and types of experience with the environment are vital in the development of the nervous system of a particular species, it is called *experience expectant*. *Experience dependent* means that the environmental input is not necessary for brain development but that it does foster synaptogenesis and refinement of existing brain structures. These processes contribute to individualization rather than species-specific brain growth.

Although culturally meaningful input could be conceived to be more important for experience-dependent processes, animal data suggest that cultural variations can play a role in caregiving experience-expectant processes as well. Most of the evidence for this assertion comes from animal studies that show how socially relevant stimuli (again, for humans, embedded in culturally meaningful routines) evoke particular neural activity. For example, Horn (1991, 1995) has shown that individual neurons in the forebrain of chicks respond specifically to objects to which the young animals have been imprinted (i.e., a moving red box or cylinder). Cynader and Frost (1999) concluded: "Obviously, these specificities were not laid by genetic information alone but by the interaction of the particular set of patterns [again in humans partly culturally prescribed] presented to the chicks . . . , and the readiness and plasticity of this specialized area of the brain at that time" (p. 168).

Similarly, Liu, Diorio, Day, Francis, and Meaney (2000) showed that variations in maternal care in the rat promote hippocampal synaptic development. In addition, neurochemical changes in the developing brain are a function of particular practices in early caregiving in both animals and humans (Shonkoff & Phillips, 2000). Specifically, the licking and grooming that a mother rat does to her pups enhances the production of serotonin and thyroid hormone, two important substances in the developing brain neurochemistry. After reviewing the evidence in this area, Shonkoff and Phillips (2000) concluded that "this evidence promises to help explain how alterations in the environment early in life may have wide ranging effects on brain development and may alter patterns of behavioral responding for children with different rearing histories" (p. 193).

Finally, studies with precocial birds demonstrate that the nature of social interaction with conspecifics plays a significant role not only in species-typical perceptual organization but also in subsequent socially relevant behaviors (Gottlieb et al., 1998). For example, manipulation of social experience with siblings can lead to different learning patterns of unfamiliar maternal calls or visual preference for a familiar maternal figure (Gottlieb, 1991, 1993; Lickliter & Gottlieb, 1985; McBride & Lickliter, 1993). These findings suggest that variation in caregiving environments can affect experience-dependent processes at the functional, biochemical, and behavioral levels.

CONCLUSIONS

The purpose of this article was to begin to outline how a developmental systems approach to the study of the role of culture on development might look. Most contemporary models of human development (i.e., Baltes, 1997; Baltes, Lindenberger, & Staudinger, 1998; Bronfenbrenner & Morris, 1998; Horowitz, 2000; Lerner, 2004; Magnusson & Stattin, 1998) acknowledge the role of both environment and biology in developmental processes, yet most of them fail to explicate in detail the potential mechanisms by which culture and biology closely operate. Two recent exceptions deserve special attention.

Kennepohl (1999) outlined what he called a *connectionist model of culture–brain interaction*, which proposes three potential mechanisms for the influence of culture on neural development and an explanation of how cultural models can be acquired by the individual brain. The three mechanisms, initially suggested by Fabregas (1982), are: (a) Ecological surroundings associated with culturally meaningful dimensions may selectively activate or tune neuronal connections; (b) cultural factors in early child learning differentially and dynamically affect brain development, particularly in the creation of neural networks; and (c) adult brains can still adapt, and acquire new languages and cultural systems, in the form of increased synaptic connections and increased neurochemical production. Kennepohl's explanation of how culture can be acquire "physicality" in the brain uses a connectionist or a parallel distributed processing model whereby cultural schemas (i.e., it is polite to wait one's turn to talk) become part of the neural circuitry of the brain. These circuitries represent the neural bases for culturally meaningful perceptual filters, behaviors, motivations, emotions, and so on, that are acquired early on and become more difficult to alter (although they remain relatively open) as development proceeds.

Another relevant model was proposed by Li (2003), who outlined what he called a *cross-level dynamic bicultural coconstructive framework of development*. In this model, a series of interconnected interactive processes (culture and context driven on one hand, and neurobiology driven on the other hand), interacting at different levels of developmental plasticity and time scales, are continuously generating and implementing biocultural, coconstructed influences on development. These influences are continuously being integrated into the individual's ontogenetic development at multiple levels of analyses. He also acknowledged the two operational spheres of culture: (a) the inherited symbolic, institutional, and historical facet and (b) the more process-oriented dimension involving day-to-day social interactions. He asserted that "bringing the collective social processing aspect of culture to the foreground makes obvious the social intersections taking place in the individuals' proximal developmental context as part of the processes mediating cultural influences on brain, cognitive and behavioral development" (p. 4). Unlike other theoretical models, in which culture appears as part of distal influences on human development, Li's model emphasizes the role of culture as proximal: "Together with behavioral, cognitive and neurobiological mechanisms, [culture] is the active 'coproducer' of behavioral, cognitive and neurobiological development during the individual lifespan ontogeny" (p. 4). This view is very compatible with the developmental systems approach to human development espoused here, where development is seen as the product of the interpenetration of biology and culture.

If this view is espoused by many theoretical models of development, and its nature specified by many recent ones, why is there such a paucity of research into the mechanisms behind the interpenetration of culture and biology? In part it is researchers' training in analyses-of-variance models in which the identification of independent, main effects is seen as the primary goal and interactions are far less frequently seen as the main hypotheses to be tested. It is also in part researchers' disciplinary-based training that makes them focus in one level of analyses (i.e., molecular) to the exclusion of others. We control rather than explore other levels and of course fail to identify true transactions across levels. Those of us who venture into various levels of analyses (i.e., biology, self, family, community), and accordingly use mixed methods that reflect high standards defined by various disciplines, are charged by the true disciplinarians as not doing "real" science.

To make breakthroughs in this important area of inquiry, methods (i.e., from ethnography to brain imaging) and theoretical frameworks from multiple disciplines (i.e., from cultural anthropology and cognitive neuroscience) will have to be mixed. It is unfortunate that, as exemplified by a recent call for action, we still have a long way to go. In the October 2001 publication of the *Observer*, published by the American Psychological Society, Snibbe asserted that "as the newest kid on psychological science's block, social cognitive neuroscience is uniquely poised to address and correct the limitations of its parents' fields ... [their conclusion] that the phenomena captured in their laboratories generalize to all human beings." She went on to assert as the solution that "a slightly more useful practice [than reporting the subjects' ethnicity] would be to control for such sociocultural variables as gender, ethnic identification and socioeconomic status." Even if identified as a theoretically important area of scientific study, the final recommendation is to do the minimum and address sociocultural influences in the phenomena as a confounding variable.

However, promise is seen in some areas of social cognitive neuroscience. The newly emergent field (Adolps, 2003; Ochsner & Lieberman, 2001) seeks to understand human functioning at three intersecting levels: (a) social, (b) cognitive, and (c) neural. Using functional neuroimaging techniques (functional magnetic resonance imaging and positron emission tomography), researchers can study the interconnection between socioemotional phenomena and related brain activity in both cortical and subcortical brain structures. So, for example, the amygdala, a subcortical structure, has been implicated in the processing of stereotyping (i.e., perception of out-group vs. in-group members) and evaluative (positive–negative) categorizations (i.e., Hart et al., 2000; Phelps, O'Connor, Cunningham, Funayama, Gatenby, Gore, & Banaji, 2000), two social processes that are theoretically linked to enculturation. However, none of the work in this emergent field has targeted as a program of research cultural processes or the documentation of the interpenetration of these processes with biology over the course of development.

Given the operational definition of *culture* as daily routines and the established impact of socially meaningful stimuli in brain functioning, a logical program of research would be to identify what aspects of cultural routines become embedded in the brain. If one presents culturally appropriate smells, for example, to other smells that have similar physical properties, does the brain respond differently? Does the same happen with culturally appropriate music, language, handling, feeding, play, and so on? Are there any subcortical and cortical areas that respond differently to culturally defined appropriate stimuli? How early in life do infants make these discriminations? Are the same brain areas involved in humans versus nonhumans?

Cautionary notes have to be made, because in this emergent field "real life human social behavior is difficult to assess, and functional imaging or invasive studies are difficult to perform" (Adolps, 2003, p. 120). But quite "complex" social phenomena, such as in-group and out-group biases, sympathy toward others, and the emotional nature of another's person gaze, have been operationalized in the laboratory (i.e., Decety & Chaminade, 2003; Phelps, Cannistraci, & Cunningham, 2003; Wicker, Perrett, Baron-Cohen, & Decety, 2003). However, the fruitfulness of interdisciplinary collaborations in this emerging field is already evident.

Acknowledging the interpenetration of culture and biology in theoretical models of development is just the initial step in this area of inquiry. The stipulation and delineation of the mechanisms by which this interpenetration occurs over the course of development are as necessary as the initial acknowledgment. In addition, as advances in the areas of social cognitive neuroscience—and potentially, the functional mapping of the genome—help us understand parts of the equation, similar efforts have to be made to operationalize cultural influences in such processes. Measurement and analyses at different levels, which involve cross-disciplinary efforts, are necessary. If not, the pendulum of the futile nature–nurture controversy will continue as the elephant in the room continues to be ignored.

ACKNOWLEDGMENTS

This work was partially supported by the Mitlemann Family Directorship at the Center for the Study of Human Development and the Edith Goldthwaite Miller Faculty Research Fellowship at the Pembroke Center, both at Brown University, and the W. T. Grant Foundation.

REFERENCES

Adolps, R. (2003). Investigating the cognitive neuroscience of social behavior. *Neuropsychologia, 41,* 119–126.

Ainsworth, M. D. S. (1967). *Infancy in Uganda: Infant care and the growth of love.* Baltimore: Johns Hopkins University Press.

Ainsworth, M. D. S., Blehar, M. C., Waters, E., & Wall, S. (1978). *Patterns of attachment: A psychological study of the strange situation.* Hillsdale, NJ: Lawrence Erlbaum Associates, Inc.

Baltes, P. B. (1997). On the incomplete architecture of human ontogeny: Selection, optimization, and compensation as foundation of developmental theory. *American Psychologist, 52,* 366–380.

Baltes, P. B., Lindenberger, U., & Staudinger, U. M. (1998). Life-span theory in developmental psychology. In W. Damon (Ed.), *Handbook of child psychology* (5th ed., Vol. 1, pp. 1029–1143). New York: Wiley.

Black, J. E., & Greenough, W. T. (1986). Introduction of pattern in neural structure by experience: Implications for cognitive development. In M. E. Lamb, A. L. Brown, & B. Rogoff (Eds.), *Advances in developmental psychology* (Vol. 4, pp. 1–50). Hillsdale, NJ: Lawrence Erlbaum Associates, Inc.

Bornstein, M. H., Azuma, H., Tamis-LeMonda, C., & Ogino, M. (1990). Mother and infant activity and interaction in Japan and in the United States: I. A comparative macroanalysis of naturalistic exchange. *International Journal of Behavioral Development, 13,* 267–287.

Bornstein, M. H., & Cote, L. R. (2001). Mother–infant interaction and acculturation: I. Behavior comparisons in Japanese American and South American families. *International Journal of Behavioral Development, 25,* 549–563.

Bowlby, J. (1969). *Attachment and loss: Vol. I. Attachment.* New York: Basic Books.

Bretherton, I. (1995). The origins of attachment theory: John Bowlby and Mary Ainsworth. In S. Goldberg, R. Muir, & J. Kerr (Eds.), *Attachment theory: Social, developmental, and clinical perspectives* (Vol. 3, pp. 45–84). Hillsdale, NJ: Analytic Press.

Bronfenbrenner, U., & Morris, P. A. (1998). The ecology of developmental processes. In W. Damon (Ed.), *Handbook of child psychology* (5th ed., Vol. 1, pp. 993–1028). New York: Wiley.

Cynader, M. S., & Frost, B. J. (1999). Mechanisms of brain development: Neuronal sculpting by the physical and social environment. In D. P. Keating & C. Hertzman (Eds.), *Developmental health and the wealth of nations* (Vol. 8, pp. 153–184). New York: Guilford.

Decety, J., & Chaminade, T. (2003). Neural correlates of feeling sympathy. *Neuropsychologia, 41,* 127–138.

Fabregas, H. (1982). Brain, culture, and neuropsychiatric illness. In I. Al-Issa (Ed.), *Culture and psychopathology* (pp. 361–385). Baltimore: University Park Press.

Fausto-Sterling, A. (1992). *Myths of gender: Biological theories about women and men.* New York: Basic Books.

Fausto-Sterling, A. (2000). *Sexing the body: Gender politics and the construction of sexuality.* New York: Basic Books.

Fausto-Sterling, A. (2003). The problem with sex/gender and nature/nurture. In S. J. Williams, L. Birke, & G. A. Bendelow (Eds.), *Debating biology: Sociological reflections on health, medicine and society* (pp. 123–132). New York: Routledge.

Fogel, A. (1993). *Developing through relationships: Origins of communication, self and culture.* London: Harvester Wheatsheaf.

García Coll, C. T. (1990). Developmental outcome of minority infants: A process oriented look into our beginnings. *Child Development, 61,* 270–289.

García Coll, C. T., Bearer, E. L., & Lerner, R. M. (Eds.). (2004). *Nature and nurture: The complex interplay of genetic and environmental influences on human behavior and development.* Mahwah, NJ: Lawrence Erlbaum Associates, Inc.

García Coll, C. T., Lamberty, G., Jenkins, R., McAdoo, H. P., Crnick, K., Wasik, B. H., & Vazquez García, H. (1996). An integrative model for the study of developmental competencies in minority children. *Child Development, 67,* 1891–1914.

García Coll, C. T., Sepkoski, C., & Lester, B. M. (1981). Cultural and biomedical correlates of neonatal behavior. *Developmental Psychobiology, 14,* 147–154.

Gottlieb, G. (1991). Social induction of malleability in ducklings. *Animal Behavior, 41,* 953–962.

Gottlieb, G. (1992). *Individual development and evolution: The genesis of novel behavior.* New York: Oxford University Press.

Gottlieb, G. (1993). Social induction of malleability in ducklings: Sensory basis and psychological mechanism. *Animal Behavior, 45,* 707–719.
Gottlieb, G. (2004). Normally occurring environmental and behavioral influences on gene activity: From central dogma to probabilistic epigenesis. In C. T. García Coll, E. L. Bearer, & R. M. Lerner (Eds.), *Nature and nurture: The complex interplay of genetic and environmental influences on human behavior and development* (pp. 85–106). Mahwah, NJ: Lawrence Erlbaum Associates, Inc.
Gottlieb, G., Wahlsten, D., & Lickliter, R. (1998). The significance of biology for human development: A developmental psychobiological systems view. In W. Damon (Ed.), *Handbook of child psychology* (5th ed., pp. 233–274). New York: Wiley.
Greenough, W. T., & Black, J. E. (1992). Induction of brain structure by experience: Substrates for cognitive development. In M. R. Gunnar & C. A. Nelson (Eds.), *Developmental behavior neuroscience* (Vol. 24, pp. 155–200). Hillsdale, NJ: Lawrence Erlbaum Associates, Inc.
Grossmann, K. E. (1995). The evolution and history of attachment research and theory. In S. Goldberg, R. Muir, & J. Kerr (Eds.), *Attachment theory: Social, developmental, and clinical perspectives* (Vol. 4, pp. 85–122). Hillsdale, NJ: The Analytic Press.
Grossmann, K., Grossmann, K. E., Spangler, G., Suess, G., & Unzer, L. (1985). Maternal sensitivity and newborns' orientation responses as related to quality of attachment in northern Germany. *Monographs of the Society for Research in Child Development, 50* (1–2, Serial No. 209), 233–256.
Hart, A. J., Whalen, P. J., Shin, L. M., McInerney, S. C., Fischer, H., & Rauch, S. L. (2000). Differential response in the human amygdala to racial outgroup vs. ingroup face stimuli. *Neuroreport, 11,* 2351–2355.
Harwood, R. L. (1992). The influence of culturally derived values on Anglo and Puerto Rican mothers' perceptions of attachment behavior. *Child Development, 63,* 822–839.
Harwood, R. L., Schoelmerich, A., Schulze, P. A., & Gonzalez, Z. (1999). Cultural differences in maternal beliefs and behaviors: A study of middle-class Anglo and Puerto Rican mother–infant pairs in four everyday situations. *Child Development, 70,* 1005–1016.
Horn, G. (1991). Imprinting and recognition memory: A review of neural mechanisms. In R. J. Andrew (Ed.), *Neural and behavioral plasticity* (pp. 219–261). Oxford, England: Oxford University Press.
Horn, G. (1995). Imprinting, or in search of the engram along the Fos way. In M. Burrows, T. Matheson, P. L. Newland, & H. Schuppe (Eds.), *Proceedings of the 4th International Congress of Neuroethology* (p. 7). Stuttgart, Germany: Thieme.
Horowitz, F. D. (2000). Child development and the PITS: Simple questions, complex answers, and developmental theory. *Child Development, 71,* 1–10.
Keating, D. P., & Hertzman, C. (1999). *Developmental health and the wealth of nations: Social, biological, and educational dynamics.* New York: Guilford.
Kennepohl, S. (1999). Toward a cultural neuropsychology: An alternative view and a preliminary model. *Brain and Cognition, 41,* 365–380.
Lerner, R. M. (1998). Theories of human development: Contemporary perspectives. In W. Damon (Ed.), *Handbook of child psychology* (5th ed., Vol. 1, pp. 1–24). New York: Wiley.
Lerner, R. M. (2004). Genes and the promotion of positive human development: Hereditarian versus developmental systems perspectives. In C. T. García Coll, E. L. Bearer, & R. M. Lerner (Eds.), *Nature and nurture: The complex interplay of genetic and environmental influences on human behavior and development* (pp. 1–33). Mahwah, NJ: Lawrence Erlbaum Associates, Inc.
Lewontin, R. C. (1982). Organism and environment. In H. C. Plotkin (Ed.), *Learning, development and culture* (pp. 151–170). New York: Wiley.
Li, S.-C. (2003). Biocultural orchestration of developmental plasticity across levels: The interplay of biology and culture in shaping the mind and behavior across the life span. *Psychological Bulletin, 129,* 171–194.

Lickliter, R., & Gottlieb, G. (1985). Social interaction with siblings is necessary for visual imprinting of species-specific maternal preferences in ducklings (*Anasplatyrhynchos*). *Journal of Comparative Psychology, 99,* 371–379.

Liu, D., Diorio, J., Day, J. C., Francis, D. D., & Meaney, M. J. (2000). Maternal care, hippocampal synaptogenesis and cognitive development in rats. *Nature Neuroscience, 8,* 799–806.

Magnusson, D., & Stattin, H. (1998). Person–context interaction theories. In W. Damon (Ed.), *Handbook of child psychology* (5th ed., Vol. 1, pp. 685–759). New York: Wiley.

McBride, T., & Lickliter, R. (1993). Visual experience with siblings fosters species-specific responsiveness to maternal visual cues in bobwhite quail chicks. *Journal of Comparative Psychology, 107,* 310–327.

Miyake, K., Chen, S., & Campos, J. (1985). Infants' temperament, mothers' mode of interaction and attachment in Japan: An interim report. *Monographs of the Society for Research in Child Development, 50*(1–2, Serial No. 209), 276–297.

Ochsner, K. N., & Lieberman, M. D. (2001). The emergence of social cognitive neuroscience. *American Psychologist, 56,* 717–734.

Overton, W. F. (1998). Developmental psychology: Philosophy, concepts, and methodology. In W. Damon (Ed.), *Handbook of child psychology* (5th ed., Vol. 1, pp. 107–188). New York: Wiley.

Oyama, S. (2000). *Evolution's eye: A systems view of the biology–culture divide*. Durham, NC: Duke University Press.

Phelps, E. A., Cannistraci, C. J., & Cunningham, W. A. (2003). Intact performance on an indirect measure of race bias following anygdala damage. *Neuropsychologia, 41,* 203–208.

Phelps, E. A., O'Connor, K. J., Cunningham, W. A., Funayama, E. S., Gatenby, J. C., Gore, J. C., & Banaji, M. (2000). Performance on indirect measures of race evaluation predicts amygdala activation. *Journal of Cognitive Neuroscience, 12,* 729–738.

Rogoff, B. (2003). *The cultural nature of human development*. New York: Oxford University Press.

Sagi, A., Lamb, M. E., Lewkowicz, K. S., Shoham, R., Dvir, R., & Estes, D. (1985). Security of infant–mother, –father, and –metapelet among kibbutz reared Israeli children. *Monographs of the Society for Research in Child Development, 50*(1–2, Serial No. 209), 257–275.

Schieffelin, B. B. (1990). *The give and take of everyday life: Language socialization of Kaluli children*. New York: Cambridge University Press.

Schieffelin, B. B., & Ochs, E. (1986). Language socialization. In B. Siegel (Ed.), *Annual review of anthropology* (pp. 163–191). Palo Alto, CA: Annual Reviews.

Shonkoff, J. P., & Phillips, D. A. (Eds.). (2000). *From neurons to neighborhoods: The science of early childhood development*. Washington, DC: National Academy Press.

Shweder, R. A., Goodnow, J., Hatano, G., LeVine, R. A., Markus, H., & Miller, P. (1998). The cultural psychology of development: One mind, many mentalities. In W. Damon (Ed.), *Handbook of child psychology* (5th ed., Vol. 1, pp. 865–938). New York: Wiley.

Snibbe, A. C. (2001). New kid should pay attention to cultural variability. In *Observer*, by American Psychological Society. Retrieved March 12, 2004. Web site address: http://www.psychologicalscience.org/observer/1001/hollywood_side.html

Spelke, E. S., & Newport, E. L. (1998). Nativism, empiricism, and the development of knowledge. In W. Damon (Ed.), *Handbook of child psychology* (5th ed., Vol. 1, pp. 275–340). New York: Wiley.

Thelen, E., & Smith, L. B. (1998). Dynamic systems theories. In W. Damon (Ed.), *Handbook of child psychology* (5th ed., Vol. 1, pp. 563–634). New York: Wiley.

Tronick, E. Z., Winn, S., & Morelli, G. A. (1985). Multiple caretaking in the context of human evolution: Why don't the Efe know the Western prescription to child care? In M. Reite & T. Field (Eds.), *The psychobiology of attachment and separation* (pp. 293–321). San Diego, CA: Academic.

van Ijzendoorn, M. H., & Kroonenberg, P. M. (1988). Cross-cultural patterns of attachment: A meta-analysis of the strange situation. *Child Development, 59,* 147–156.

Weisner, T. S. (2002). Ecocultural understanding of children's developmental pathways. *Human Development, 45,* 275–281.

Werker, J., & Lalonde, C. (1988). Cross-language speech perception: Initial capabilities and developmental change. *Developmental Psychology, 24,* 672–683.

Werker, J., & Tees, R. (1984). Cross-language speech perception: Evidence for perceptual reorganization during the first year of life. *Infant Behavior and Development, 7,* 49–63.

Wicker, B., Perrett, D., Baron-Cohen, S., & Decety, J. (2003). Being the target of another's emotion: A PET study. *Neuropsychologia, 41,* 139–146.

Whose Lives? How History, Societies, and Institutions Define and Shape Life Courses

Karl Ulrich Mayer
Yale University

This article outlines how current sociology constructs life courses. First, a set of general heuristics is provided. Second, the development of life course sociology over the last 50 years is traced as an intellectual process whereby the *life course* has emerged as an analytical construct in addition to such concepts as human development, biography, and aging. A differential life course sociology has gradually developed in which contexts are specified according to time and place. Third, these differential constraints operating on life courses are illustrated from the perspective of 2 research areas. One perspective introduces historical periods as a sequence of regimes that regulate life courses. Another perspective looks at cross-national differences and especially focuses on institutions as the mechanisms by which life courses are shaped. The article concludes with reflections about the relation between the variable social contexts of life courses and human development.

In recent years, there has been a marked shift in the way human development and human life courses are being perceived. Infants and children are seen as producers, or at least as coproducers, of their own development (Lerner & Busch-Rossnagel, 1981). Parent–child relationships and socialization processes are categorized much less as one-way streets where parents and other socialization agents imprint and impose their values and habits on children and adolescents but rather as areas of mutual interaction where it remains open who influences whom more (Krappmann, 2001; Kreppner, 1999). The old idea that teachers effectively transfer knowledge and character has given way to unending reports about unruly classes and resistant pupils. Sociologists have newly celebrated the significance of human agency (Giddens, 1984) and the individualization of life decisions and

---

Requests for reprints should be sent to Karl Ulrich Mayer, Max Planck Institute for Human Development, Center for Sociology and the Study of the Life Course, Lentzeallee 94, D-1495 Berlin, Germany. E-mail: Sekmayer@mpib-berlin.mpg.de

lifestyles in patchwork biographies (Beck, 1986). Fewer daily working hours, coupled with considerable disposable income, open up a variety of self-chosen milieus and habitus. Ever earlier onsets and ever later conclusions of adolescence and transitions to adulthood are being interpreted as significant extensions of personal autonomy: "getting into one's own" (Modell, 1991).

In contrast, development regulated by universal biological principles of maturation and decline (Piaget, 1970); the harsh discipline of families, workplaces, and a variety of other institutions (Foucault, 1977); and life courses determined and constrained by tradition, collective class fate (Thompson, 1976), or the whims of historical catastrophes such as the Great Depression (Elder, 1974), the world wars (Winter, 1986), and the Holocaust (Kertesz, 1992; Levi, 1995) seem to echo pictures of an old past. If at all, it appears as if it is the lack of limits of options, the unlimited flexibilization (Sennett, 2000) and pluralization that pose the postmodern condition.

Under such premises it seems almost odd to raise the question of how life courses are shaped by forces external to the individual person, how historical conditions, the good or bad fortunes of national citizenship or institutional arrangements built the tracks that individual trajectories are bound to follow (Mayer, 1986). Three lines of argument can be used to defend taking up such seemingly outmoded perspectives. One line of argument goes back to Immanuel Kant, who insisted in his philosophy of the mind that determinism and autonomy, constraint and choice, are regulative principles of potential knowledge and moral behavior that do not rule each other out but rather constitute different and mutually exclusive modalities of how to view the world. A second line of argument reminds us that "individualism" and its opposites are in themselves historically variable sociocultural constructs (Meyer, 1986). The relative extent to which we perceive the person and his or her life as actors with their own scripts is a matter of culturally pre-fixed lenses. Thus, whether it is at all possible to resolve the issue of growing or declining personal autonomy, of relative degrees of choice and constraint, is open to debate. The third line of argument insists not only that lives in countries less fortunate than the G-7 club are to a much higher extent bound by the arbitrariness of the social class and national citizenship into which one is born but that also after the exceptional postwar periods of relative affluence, constraints and dependency are on the rise.

In this article I outline how the social construction of life courses is currently being defined and developed in sociology. In a first step, I summarize my version of a general theory of the social organization of the life course. In a second step, I portray the development of life course sociology over the last 50 years as an intellectual process whereby the *life course* has emerged as an analytical construct in addition to and separate from such notions and terms as *human development, biography,* and *aging.* Moreover, general ideas of how human lives are shaped by social and historical circumstances have been gradually replaced by a kind of differential life course sociology in which contexts are specified according to time and place. After these two introductory stepping stones, I look at the social constraints operating on life courses from the perspective of two different research areas. The first perspective introduces historical periods as a temporal sequence of contextual regimes that regulate life courses. The second perspective looks at cross-national differences in patterns of life courses and especially focuses on institutions as the mechanisms by which lives are channeled in specific ways. Here I argue that the development of life course sociology has benefited greatly from recent advances in the fields of comparative welfare state research (Leisering & Leibfried, 1999) and from the research on "varieties of capitalism" (Hall & Soskice, 2001). I conclude the article with some reflections about the relation between the variable social contexts of life courses and human development.

My aim is not to provide a systematic review that would do full justice to the literature and state of empirical research in this field but rather to present arguments in a more exemplary manner. I rely heavily on my own material (and that of my colleagues) and evidence from the nine cohort surveys of the German Life History Study (Brückner & Mayer, 1998), and I borrow liberally from a wide array of publications that originated in the Center for Sociology and the Study of the Life Course at the Max Planck Institute for Human Development, which I have directed since 1983. Although I heavily rely on work from my associates in the German Life History Study, they share no responsibility for the edifice I am going to build upon it.

THE LIFE COURSE FROM THE PERSPECTIVE OF SOCIOLOGY

With the term *life course* sociologists denote the sequence of activities or states and events in various life domains spanning from birth to death.[1] The life course is thus seen as the embedding of individual lives into social structures primarily in the form of their partaking in social positions and roles, that is, in regard to their membership in institutional orders. The sociological study of the life course, therefore, aims at mapping, describing, and explaining the synchronic and diachronic distribution of individual persons into social positions across the lifetime. One major aspect of life courses is their internal temporal ordering, that is, the relative duration times in given states as well as the age distributions at various events or transitions.

---

[1] This section is a revised and shortened version of Mayer, 2003, pp. 464–468. © Copyright 2003. Used by permission of Kluwer Academic/Plenum Publishers.

How do order and regularities in life courses come about? Sociologists primarily look for three mechanisms to account for the form and outcomes of life courses. The first mechanism is the degree and kind to which societies are internally differentiated into subsystems or institutional fields (Mayer & Müller, 1986). This is often taken to be the most obvious and important mechanism. The second mechanism lies in the internal dynamic of individual lives in group contexts. Here, one searches for conditions of behavioral outcomes in the prior life history or in norm-guided or rationally purposive action. The third mechanism derives from the basic fact that it is not simply society on the one hand and the individual on the other that are related to each other, but aggregates of individuals in the form of populations such as birth cohorts or labor market entry cohorts (Mayer & Huinink, 1990).

I now illustrate each of these three life course mechanisms in turn. How do institutions corresponding to various subsystems shape life courses? The educational system defines and regulates educational careers by its age-graded and time-scheduled sequences of classes; its school types and tracks; and its institutions of vocational and professional training and higher learning, with their hierarchical and time-related sequence of courses and certificates. Labor law defines who is gainfully employed and who is unemployed or out of the labor force and, thus, employment trajectories. The occupational structure defines careers by conventional or institutionalized occupational activities, employment statuses and qualification groups, segmentation, and segregation. The supply of labor determines the opportunity structure and, thus, the likelihood of gaining entry into an occupational group or of change between occupations and industrial sectors. Firms provide by their internal functional and hierarchical division of labor career ladders and the boundaries for job shifts between firms and enterprises. In a similar manner, the institutions of social insurance and public welfare define the status of being ill, the duration of maternity leave, the age or employment duration until retirement, and so on. Family norms and laws constitute the boundaries between being single or in nonmarital unions, married, and divorced. Finally, the spatial structure of societies, as well as forms of property, define the interaction with family roles and forms of household trajectories of residential mobility, household changes, and migration.

The second mechanism for shaping life courses focuses on life trajectories and their precedents. Descriptively, research tends to concentrate on transition or hazard rates, that is, the instantaneous rates at which a well-defined population at risk makes certain transitions, for example, into first employment, first motherhood, retirement, and so on, within given time intervals. The explanatory question for life course research, then, is whether certain life course outcomes are shaped not only by situational, personal, or contextual conditions but also by experiences and resources acquired at earlier stages of the biography such as incomplete families in childhood (Grundmann, 1992), prior job shifts (Mayer, Diewald, & Solga,

1999), prior episodes of unemployment (Bender, Konietzka, & Sopp, 2000), educational careers (Henz, 1996), or vocational training and early career patterns (Hillmert, 2001a; Konietzka, 1999; Solga, 2003).

There is one important additional point to be made in this context. Looking for causal mechanisms on the micro level of the individual biography does not resolve the issue of whether the individual is more of an active agent or more of a passive object in the processes that shape the life course or—to put it in different terms—whether selection or adaptation by choice is of primary importance (Diewald, 1999, chap. 2; Nollmann, 2003). Sociologists tend to be split on this issue. Some would emphasize cultural scripts, some would stress social norms, and others would bet on rational choice. On the whole, however, sociologists tend to believe more in selection than in choice. First of all, already the institutional contexts as described earlier narrow down to a large extent which life avenues are open and which are closed. Second, within given institutional contexts, individuals are probably more frequently being selected than selecting themselves. This is related to another sociological axiom: If material resources, power, and authority, as well as information and symbolic goods, are distributed very unequally within given societies, then it follows that more people have to accommodate than have the opportunity to exert control.

The third mechanism that one can look for in unraveling patterns in life courses has to do with the fact that it is not single individuals but populations that are allocated to, and streamlined through, the institutional fabric of society across the lifetime—for example, the size of one's cohort, as well as the preceding and succeeding cohorts, influences individuals' opportunities way beyond individual or situational conditions (Hillmert, 2001b; Ryder, 1965, 1980). Similarly, the dynamics of union formation and marriage through which one's own chances to find a partner are shaped change over time depending on the behavior of others searching at the same time (Hernes, 1972, 1976).

From the perspective of sociology, then, life courses are considered not as life histories of persons as individuals but as patterned dynamic expressions of social structure. These dynamics operate in populations or subsets of populations, are governed intentionally or unintentionally by institutions, and are the intentional or unintentional outcomes of the behavior of actors. Patterns of life courses are, however, not only products of societies and a part and parcel of social structure but also important mechanisms for generating social structures as the aggregate outcome of individual steps throughout the life course. One transparent example of these processes is evident in the fact that the age and cohort structure of a population is the highly consequential result of a multitude of fertility behaviors and decisions. Likewise, the employment structure is the outcome of a multitude of individual employment trajectories.

Finally, the relation to historical time is crucial for the sociological study of life courses because life courses are embedded in definite strands of historical periods as well as in the collective life history of families and birth cohorts. Life courses are subject not only to historical circumstances at any time but also to the cumulative or delayed effects of earlier historical times on the individual life history or the collective life history of birth cohorts (or marriage cohorts or employment entry cohorts).

Our heuristic for the study of life courses is thus guided by four signposts (Huinink, 1995; Mayer & Huinink, 1990): First, individual life courses are to be seen as part and product of a societal and historical multilevel process. They are closely tied to the life courses of other people (parents, partners, children, work colleagues, etc.) and the dynamics of the social groups of which they are a member. They are highly structured by social institutions and organizations and their temporal dynamics. Second, the life course is multidimensional. It develops in different mutually related and mutually influencing life domains, such as work and the family. It also unfolds in the context of biological and psychological maturation and decline. Third, the life course is a self-referential process. The person acts or behaves on the basis of prior experiences and resources. Therefore, one must expect endogenous causation already on the individual level. Via aggregation this then also becomes true for the collective life course of birth cohorts or generations. Individuals' and generations' pasts facilitate and constrain their futures. This is the meaning of the phrase *die Gleichzeitigkeit des Ungleichzeitigen* (the contemporaneity of the uncontemporaneous) characterizing the interdependency of generations. The various age groups live together in a common present, but each brings to it its own particular past. Fourth, through the manner in which people live and construct their own individual lives, they reproduce and change social structures. This can either happen through "simple" aggregation processes or through immediate or intermediate institution formation. An example of the latter would be that a growing proportion of fully employed mothers exert electoral pressure to change schools into institutions that take care of children for most of the day.

One might also ask in which sources sociologists of the life course expect the greatest share of variance in life course outcomes to be explained. The largest part of variation will usually be expected to reside in those external structures within society that are closely tied to the division of labor, that is, the occupational structure and the structure of employment in various industrial sectors and the educational systems. The reason for this is that both the distribution of initial resources, of resulting income rewards, and the distribution of positions that form the basic opportunity structure and into which people are sorted, are intimately tied to these institutional fields. Thus, life course patterns are expected to vary greatly across social classes or status groups (Mayer & Carroll, 1990). The second-largest source of variation sociologists tend to locate would be in the division of labor within households, that is, the way women and men in families and other unions allocate their lifetimes for economic and family roles (Ben-Porath, 1979; Sørensen, 1990). The third important source of variation life course sociologists would look for relates to the differential intervention of the state in the form of the modern welfare state (Huinink et al., 1995; Leisering, 2003; Mayer & Müller, 1986). It is, therefore, the so-called *welfare mix* (i.e., the relative importance and manner of interconnectedness of economic markets, the family, and the state across historical time and across contemporary societies) that sociologists see as the major determinant of life course patterns (Esping-Andersen, 1999).

THE ARCHAEOLOGY OF LIFE COURSE SOCIOLOGY

Life course sociology emerged and developed over several decades (see Figure 1). The first stage coincided with the years between the two world wars. Theoretical notions of development and the life cycle as proposed by psychologists such as Charlotte Bühler were not clearly separated from the methodological instrument of life histories (Thomas & Znaniecki, 1918) designed to capture personality development, social conditions, and historical context at the same time. In the same period, Karl Mannheim (1952) proposed another very synthetic concept—the generation—that fused quite general ideas about the social metabolism (i.e.,

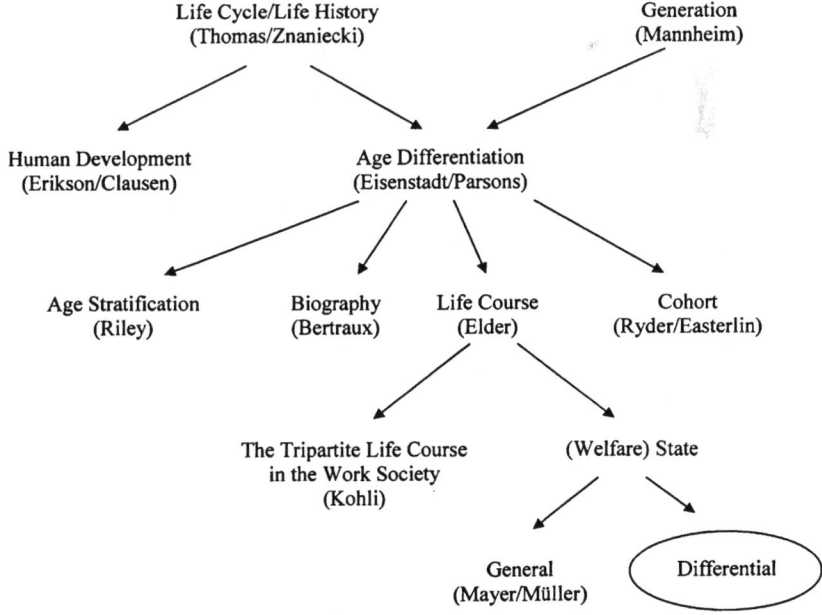

FIGURE 1 The "archaeology" of comparative life course sociology.

social change via the succession of cohorts) with ideas about historical styles and historically specific collective actors.

During the second stage—in the 1940s and 1950s—the more psychological traditions of human development (Clausen, 1986; Erikson, 1980) focusing on internal personal dynamics in mostly group contexts became more clearly distinguished from the sociological concept of age differentiation (Eisenstadt, 1964; Parsons, 1942) as a structural category. It should be stressed, however, that the close link among psychological, social psychological, social, and historical perspectives remained a major focus, as demonstrated, for instance, in the extensive work of Glen Elder (1974, 2001).

During the third stage—in the 1960s and 1970s—the broader concept of age differentiation was further subdivided by:

1. The narrower concept of age stratification (Riley, Kahn, & Foner, 1994), which stressed not only functional specificity but also inequalities in resource allocation and power.
2. Biography as subjective narrative (Bertaux, 1981; Kohli, 1981).
3. Generation as a cultural construct (Bude, 1995).
4. The life course as the social organization of lives.
5. The demographic concept of the cohort (Ryder, 1965, 1980).

It is worth noting, however, that in almost all of these attempts at concept formation and theory building the major focus was still on the development of fairly broad universal and general notions. Personal dynamics were now more clearly seen in contrast to diachronic social contexts and historical experiences, and the quest for subjective meaning in life designs and life reviews was pitted against the objectivity of demographic accounts on collective cohorts. Only very slowly, and under the pervasive influence of social historians such as Aries (1973), Hareven (1986, 1996), and Modell (1991; Modell, Furstenberg, & Hershberg, 1976), did it become possible to have the knowledge about the variability in the social and cultural organization of life courses come to the forefront.

During the fourth stage—in the 1970s and 1980s—there were several attempts to pinpoint the specificity of life courses (and biographies) both within and in contrast to past societies. On the one hand, Kohli (1985) and others tried to demonstrate how life courses derive from the prerequisites of (capitalist) economies where lives and life stages center around work. On the other hand, the uniqueness of modern life courses was derived from the emergence of the welfare state (Mayer & Müller, 1986; Mayer & Schoepflin, 1989). Even during this stage, however, "the work society" versus "the welfare state" and "modern" life courses versus "traditional" life courses were the focus of the debate rather than related to cross-national and historical variation.

It was only in the middle half of the 1980s and 1990s that something like a "differential" life course sociology developed, that is, descriptions of how patterns of life courses varied between more and more delimited historical periods and between societies. Although rough dichotomies, such as traditional versus standardized life courses or open versus closed societies, were used at first, gradually more institutional specifics were marshalled. One early case of the latter was the attempt to relate processes of job mobility to the institutional variation in education and training systems (Allmendinger, 1989b). The more detailed the supportive evidence was, the closer one also moved to the question of what accounted for the observed differences. It is my thesis that the development of historical and cross-national comparative life course research opened up the opportunity to come to grips with the mechanisms that might explain how social contexts shape individual life courses.

LIFE COURSES ACROSS (RECENT) HISTORY

That human lives are embedded in social contexts and are powerfully regulated and constrained by such contexts can be made plausible especially well by historical comparisons. On the one hand, there is the universal human condition handed down by an evolutionary process spanning many thousands of years. In this respect one can easily recognize that maturation and functional decline; sexual and material reproduction; and social hierarchies based on age, gender, and collective symbols form an underlying constant in shaping life courses (Linton, 1945). On the other hand, it is hard to underestimate the huge difference between the short life spans in early hunter/farmer societies lived in almost completely natural environments and the long life spans in present-day societies lived in mostly man-made physical and institutional environments. Although such comparisons across history are able to demonstrate the dependency of each individual life on its varying social context, they are also highly suggestive about the existence of long-term trends in the direction both toward greater diversity within populations at a given historical period and toward ever-greater autonomy and control of individual actors in how they live out their own lives.

Of course, there are also dangers and drawbacks in historical comparisons of life course patterns. One danger is that one is constructing images of the past that appear much tighter and orderly than it probably was and how it appeared to the contemporaries. Another drawback is that historical periods, however one defines them, do not usually coincide with the life spans of people and generations. The closer one gets to the present, and the smaller one's time unit of historical periods becomes, the more often will individual lives actually spread across several periods. In which sense, then, could one postulate that the social conditions of given periods "shape" life courses (Mayer & Hillmert, 2003)?

There exists by now a stylized history of the development of life courses which—in different versions—has been well described by, among others, Modell et al. (1976), Buchmann (1989), Anderson (1985), Hareven (1986), and Myles (1990, 1993). Life courses are said to have developed from a traditional/pre-industrial type, to an early and late industrial type, and after that to a postindustrial type; from the Fordist to the post-Fordist life cycle; from the standardized to the destandardized life course (see Table 1). Next, I briefly sketch and summarize such attempts. Let me stress, however, that the historical periods depicted must be taken as theoretical constructs rather than validated empirical generalizations.[2]

Under the traditional, pre-industrial life course regime, life centered around the family household and its collective survival. Schooling was nonexistent or short (only in winter when children were not needed on the farm), and training was part of family socialization in one's own or other families as servants. Marriage was delayed until either the family farm could be inherited or a farm heiress could be married off or until a sufficient stock of assets could be assembled to establish a household, build a house, lease some land, and so on. Life was unpredictable because of the vicissitudes of nature in harvests and the probability of sickness and early death (especially for women in childbirth). Economic dependency and debts were widespread. The subjective counterpart of such a life course regime was a collective rather than individual identity, fatalism, and religious complacency.

The early industrial life course regime is well captured in Rowntree's (1901/1914) image of a life cycle of poverty where industrial workers could only for a short time in their life rise above poverty when the family was still small and physical working capacity at its peak. Schooling was compulsory but ended at a relatively early age. Dependent work started with ages 12–14 and ended only with physical disability in old age. Marriage was delayed until sufficient resources for establishing a household (furniture, dowry) were accumulated and until employers were prepared to pay a family wage. Unemployment was frequent.

The next stage is postulated to be the industrial, Fordist life course regime. It is characterized by distinct life phases: schooling, training, employment and retirement, stable employment contracts, and long work lives in the same occupation and firm. A living wage for the male breadwinner could allow women to stay at home after marriage. The risks of sickness, unemployment, disability, and old age were covered and softened by an ever-more comprising system of social insurances. Age at marriage and first birth decreased into the early 1920s. Families could accumulate savings to buy their own house or apartment, and wages were age graded. Real incomes and purchasing power increased for a good part of the working life and then stabilized until retirement, when pensions and low rents or mortgage payments ensured a standard of living comparable to the one of the ac-

---

[2]The following section is a revised and shortened version of Mayer, 2001, pp. 92–97. © Copyright 2001. Used by permission of Routledge.

TABLE 1
Historical Changes in Life Course Regimes

Life Course Regimes	Traditional (ca. 1900)	Industrial (1900–1955)	Fordist/Welfare State (1955–1973)	Post-Fordist/Postindustrial (1973–Present)
Unit	Family farm/firm	Wage earner	Male breadwinner, nuclear family	Individual
Temporal organization	Unstable, unpredictable discontinuity	Life cycle of poverty, discontinuity	Standardized, stabilized, continuity, progression	Destandardized discontinuity
Education	Minimal elementary	Medium compulsory	Expansion of secondary and tertiary education and of vocational training	Prolonged, interrupted, lifelong learning
Work	Personal dependency; family division of labor	Wage relation, firm paternalism, unemployment	Full lifelong employment, upward mobility, income progression	Delayed entry, high mobility between firms and between occupations, flat income trajectories, unemployment
Family	Partial and delayed marriage; instability due to death; property centered, high fertility; early death	Delayed universal; fertility decline	Early universal marriage, early childbearing, medium fertility	Delayed and partial marriage, pluralized family forms, low fertility, high divorce rate, sequential promiscuity
Retirement/old age	With physical disability, old age dependency, early death	Regulatory or by disability, low pensions	Regulatory, medium pensions	Early retirement, decreasing pensions, increasing longevity, increasing chronic illness

Note. Reprinted from "The Paradox of Global Social Change and National Path Dependencies: Life Course Patterns in Advanced Societies" by K. U. Mayer. In A. E. Woodward & M. Kohli (Eds.), *Inclusions and Exclusions in European Societies* (pp. 89–110). Published by Routledge. Copyright © 2001. Used by permission.

tive years. Relative affluence allowed children to receive more education and training than the parental generation, and parents could afford to support their children in buying their own homes.

The subjective hallmark of such a life course regime was material progression and accumulation but also conformity to given roles within the economy and the family. Its logic followed the logic of division of labor within the nuclear family and of the family welfare as a joint utility function of the family members. Social identities were well defined and stable. Stratification homogenized, and workers were integrated into, society socially, economically, and politically.

The standardized linear and homogeneous life course that emerged in post–World War II society is generally attributed to the coming together of two forces: (a) Fordist industrial mass production, in which a relatively secure working class earning a moderate wage became established as the "universal" class, and (b) the welfare state's guarantee of income across the entire life cycle of the family. The standardization of the life course meant in a sense that workers' life chances became "middle class."

The postindustrial, post-Fordist life course regime, in contrast, can be characterized by increasing destandardization across the lifetime and increasing differentiation and heterogeneity across the population. Education has expanded in level and duration; vocational and professional training, as well as further training, have proliferated. A number of life transitions have been delayed, prolonged, and increased in age variance, and the degree of universality and of sequential orderliness has decreased. Entry into employment has become more precarious; first work contracts are often temporary; and employment interruptions due to unemployment, resumed education or training, or other times out of the labor force have increased. The rate of job shifts increase and occupations are increasingly not lifelong. Careers become highly contingent on the economic fates of the employing firms; therefore, heterogeneity across working lives increases. Downward career mobility increases relative to upward career opportunities. Working lives shorten because of later entry and frequent forced early retirement. The experience of unemployment becomes widespread but is concentrated in women, foreign workers, young people, and older workers. Age at marriage has increased. Nonmarital unions exploded and became a normal phase before marriage. Parenthood is delayed and for a significant number of couples never comes about. Divorce increases as has the number of children growing up in a single household and/or without a father present in the household. Women overtake men in their share of general education and greatly increase their occupational qualifications. Women want to work lifelong, and they have to work to augment the family budget or support themselves as single mothers. The standard of living in old age is threatened by reduced pension entitlements. The relation between the home and the working place is changing rapidly. Women are out of the house most of the day.

The subjective counterpart is *hedonistic individualism*, whereby all persons—even within a family—have their own life designs and life projects or, rather, follow egotistically the shifting material incentives and consumption idols from situation to situation.

A question then arises: Which institutional configurations shaped these various life course regimes? The following is a preliminary suggestive list:

- The traditional life course regime was regulated by the demographics of high mortality and high fertility, by prerequisites and vicissitudes of a rural economy without the benefits of the agrochemical fertilization of soil and scientific animal husbandry.
- The early industrial life course regime was subjected to an untamed capitalist economy with a weak labor movement and—because of the first demographic transition—a high supply of labor.
- The late industrial life course regime was made possible by effective coordination between capital and labor, mass production and mass consumption, macroeconomic policy intervention stabilizing economic cycles, full employment, rising real wages and standards of living, and, finally, welfare state expansion.
- For the postindustrial, post-Fordist life course regime or life course disorder, a manifold of culprits have been named: educational expansion and its unintended effects, the women's movement, value changes, individualization and self-direction, weakness of trade unions, de-industrialization, the labor market crises with spiraling structural unemployment, globalization of economic markets, and the demographic crunch produced by the low levels of fertility and decreasing mortality.

The various attempts to develop empirical accounts of historical changes in life course regimes provide strong evidence for the high degree of context-boundedness of life courses. At the same time, such historical comparisons suffer from the same weakness as the parallel and related tradition of intercohort comparisons (Mayer, 1995; Mayer & Huinink, 1993; Modell et al., 1976). "Cohorts" or "historical periods" mark differences, but the assumptions as to what causes such differences remain mostly foggy. The holistic assumption of overall regulation regimes resulting in specific patterns of life course outcomes such as "Fordism" and "post-Fordism" is more postulated than proven (Boyer & Durand, 2001; Mayer & Hillmert, 2003; Myles, 1993). Moreover, it is apparent that the assumed trends or period differences can claim little general validity as to specific timing, turning points, and direction. Cross-national comparisons and intracountry developments promise remedies in both respects and could facilitate an understanding of the mechanisms bringing about varying patterns of life course outcomes.

CROSS-NATIONAL DIFFERENCES IN LIFE COURSE REGIMES

For illustrative purposes, I now consider four categories of countries (for a more detailed discussion, see Breen & Buchmann, 2002; DiPrete, 2002; Hillmert, 2002; Leisering, 2003; Leisering & Leibfried, 2001; Mayer, 2001; Mills & Blossfeld, 2001): (a) the Scandinavian social democratic welfare states, (b) the liberal market economies, (c) the continental conservative welfare societies, and (d) the southern European countries. These are the "three Western worlds of welfare capitalism" depicted by Gøsta Esping-Andersen (1990), augmented by Italy and Spain (see also Esping-Andersen, 1999).[3]

Liberal market societies are characterized by the following institutional configuration: little stratification in the school system; no well-developed institutions for vocational training; a poor performance in training the workforce for skilled labor; high labor market flexibility in reallocating workers between firms; a relatively low level of welfare income redistribution; low social insurance levels; citizenship-based and targeted, means-tested social provisions; and poor family services.

The conservative welfare state, of which Germany is the major exemplar, caters institutions that make for stratified and selective schooling; a well-developed training system; a good performance in skill formation, and therefore high internal labor market flexibility; but highly segregated, segmented, and rigid labor markets. Social insurances are generous in comparison and are based on entitlements derived from employment. Family services are relatively poor and therefore make it difficult for women to maintain continuous work careers.

The social democratic welfare states do not stratify and segregate their secondary school system and are relatively efficient in providing vocational training within the school system. Their citizens enjoy very high levels of social insurances based on universal citizenship rights and the general tax base. Family services are excellent and therefore allow women to become fully integrated into the labor market, not least in the family services themselves.

Finally, the southern European familistic residual welfare states have stratified schooling systems, firm-based vocational training, low transfers except for pensions, and high labor market rigidity.

What are the consequences of these four different institutional configurations and political economies for the predominant life course regimes?

To capture essential aspects of such life course regimes (see Table 2), I have selected nine aspects:

---

[3]The following section is a revised and shortened version of Mayer, 2001, pp. 97–107. © Copyright 2001. Used by permission of Routledge.

TABLE 2
Political Economies and Life Course Outcomes

Outcome	Liberal	Conservative	Social Democratic	Familistic
Leaving home	Early, high variance	Medium, high variance	Early, low variance	Late, high variance
Age leaving school/training	Medium, homogeneous	High stratified	Medium	Low stratified
Labor market entry	Early, stopgap, low skill	Late, integrated, high skill	Early, integrated	Late, marginal
Firm shifts	High	Low	Low	Low
Occupational shifts	High	Low	High	Low
Income trajectories	Flat, high variance	Progressive, low variance	Flat, low variance	Progressive, high variance
Careers of women	High participation, medium level, continuous	Medium participation, low level, interrupted	High participation, low level, continuous	Low participation, low level, dualist
Family	Unstable, medium fertility	Stable, low fertility	Stable, medium fertility	Stable, low fertility
Retirement	Low replacement, high variance	High replacement, low variance, early	High replacement, low variance, late	Late, high variance

Note. Reprinted from "The Paradox of Global Social Change and National Path Dependencies: Life Course Patterns in Advanced Societies" by K. U. Mayer. In A. E. Woodward & M. Kohli (Eds.), *Inclusions and Exclusions in European Societies* (pp. 89–110). Published by Routledge. Copyright © 2001. Used by permission.

- leaving the parental home,
- the age of leaving school or formalized training,
- the process of labor market entry,
- the rate of fluctuation as work life mobility between firms,
- the rate of work life mobility between occupations,
- the shape and distribution of income trajectories,
- the degree of career involvement of women,
- fertility and the stability of families,
- median and dispersion of age at retirement.

Universal and comprehensive schooling without institutionalized apprenticeships in the liberal market societies make for a fairly standardized age at leaving secondary school around age 17. There is some variance due to high school dropouts in the United States and due to the differences between O- and

A-levels in Britain, but in comparison nonselective, comprehensive school systems standardize the length and finishing age of the formative period (Hillmert, 2001a). Labor market entry comes early even for college graduates, but the transition between education and full labor market integration is often marked by a sequence of stopgap jobs (Allmendinger, 1989a, 1989b; Oppenheimer & Kalmijn, 1995). Low-paid and marginal employment, as well as unemployment, is widespread among young workers. Educational certificates are of minor importance and occupational identities are weak, and therefore work lives are primarily structured by individual attempts to make good earnings. Commitment to given firms is low, and job shifts between firms are frequent. Deregulated labor markets foster employment but depress and polarize wages. Mean income trajectories are fairly flat across working lives because efficiency wages and seniority premiums are weak and effects of the business cycle are stronger than age effects. Labor income inequality is high, but the stability of relative income positions across the working life is low. Employment opportunities for women are good, and employment trajectories are highly continuous, but women's work is hardly optional because their share in the family budget is badly needed. Therefore, women's full-time work is the standard rather than part-time work (Blossfeld & Hakim, 1997). Probably because of the relative economic independence of women, divorce rates are high, but so are remarriage rates of women with children, who could hardly cope otherwise. Nonetheless, and despite bad family allowances and services, fertility among these countries is not at the lowest. At retirement, the replacement rate of pension income compared with the final wages is relatively low, and there is a high variance of the median age at retirement, because on the one hand, older workers can be fired easily, and on the other hand, older workers continue to work even at lower wages because of the low level of expected pension income. The major risks in this life course regime include low skills, low wages, and being one of the working poor below or close to the poverty level. For a considerable proportion of the people, the threat of a cumulative cycle of disadvantage is very real.

Conservative, corporatist welfare states stratify school and training tracks and thus induce a higher variance in the ages at which young adults leave the formative period. A prolonged educational period also pushes the age of leaving home upward, but its variance is also tied to educational and training decisions. To the extent to which training is also organized within firms, transitions to employment are smoother and integrated along the lines of occupational tracks. Training investments by both firms and young people are high, and therefore the attainment and the later use of certified skills play a large role in young people's lives. Job shifts between firms are rare, and changes between fields of occupational activities are even rarer. For those who successfully manage their labor market entry, mean income trajectories are progressive up to the early 40s and then flatten out.

Efficiency wages and seniority schemes are widespread, even in the private sector. The industrywide binding character of collective agreements and informal wage coordination between industry unions ensures relatively low degrees of wage inequality. Labor market rigidities go hand in hand with high rates of unemployment, especially for younger workers of foreign descent, women, and older workers who become laid off. Although the labor force participation of women has been increasing rapidly, the opportunities and commitments for married women with younger children are greatly limited. Career interruptions in the early years after childbirth and later part-time work are normatively expected and institutionally supported by restricted child-care and child leave options. Marriages are comparatively stable, but fertility is low. Especially for women with higher education, a dualistic behavior pattern is observable: either high career commitment with no children or career withdrawal and two children. Retirement comes early, because firms try to get rid of older workers with higher wages, but this is increasingly limited by tighter disability and old-age pension rules. The major life course risks in this political economy are long-term unemployment and being pushed into the group of labor market outsiders.

Life courses in the Scandinavian welfare states are distinct, especially in the following regard: the full-time, full working life integration of women into the labor force, a somewhat higher level of fertility until the 1990s, the nontemporary character of nonmarital unions, and the public support for people who are out of the labor force and for their reintegration (with the result of early leaving of the parental home), and, finally, relatively late ages at retirement. The major life course risks are the transitions from comprehensive school to employment with now-high levels of youth unemployment or enrollment in employment policy measures and the entrapment into low-wage, low-skill employment in the public sector for women. There is then the risk of "welfare careers" both inside and outside the employment system.

Life courses in the residual welfare states of southern Europe are for men and women closely tied to the fortunes of the larger family. Not only is access to employment highly dependent on family and kinship connections, but also most of the welfare burdens are put on families. The large number of unemployed or marginally employed young people live with their families longer than anywhere in Europe, and the care for chronically ill old people is left almost exclusively to families, with the Filipina-in-residence being the richer families' solution to old age care. Women have caught up with men in their educational attainments and occupational qualifications, and many of these qualified women delay or renounce childbearing to escape the burdens connected with motherhood. In family patterns, marriage is still almost universal, but childbearing diverges greatly between north and south, city and country, and according to the educational level of women. As a consequence, inequalities between families are high, and for individuals they tend to be cumulative across the life course. Only in regard to pensions for certain occupational groups does the southern welfare state show surprising generosity, partly in level, but especially in regard to the early age of, eligibility.

One can try to summarize these life course regimes along four dimensions:

- Which is the action unit around which life courses are primarily organized?
- What is the predominant temporal organization of states and events across the lifetime?
- How heterogeneous and unequal are life courses between social classes and between men and women?
- How do inequalities within birth cohorts develop across their collective lifetime?

In liberal market societies, the basic unit and actor in life courses is the individual. The organization of life time is not well standardized and exhibits a fair degree of discontinuity. Income inequalities in a cross-sectional perspective are both high and unstable. They are accentuated by highly unequal and dualistic access to social insurance. Individuals who can afford private insurance are well covered, and those who cannot afford private insurance are at risk of falling into poverty. The high labor market integration of women, in contrast, tends to favor equality between men and women. The relative income position across the life course is quite unstable but still tends to result in cumulative cycles of privilege and disadvantage and thus an increasing inequality across the life course.

Conservative welfare societies organize life courses around the nuclear family, although with increasing shares of life time spent outside conventional families. In comparison, life courses are still highly continuous and standardized. Cross-sectional inequalities are in the medium range and fairly stable across work life and retirement. Inequalities, however, increase between individuals who are integrated into the highly protected labor market and those who either have a hard time entering or are being phased out into early retirement via temporary unemployment or are being kept out (at least partially in life time and in working hours), as in the case of women. Some of these outsiders are cushioned by social wages and others by their families. Gender inequalities decrease somewhat: most in general education; less in occupational training, tertiary education, and employment; and much less in occupational careers. These gains are threatened, though, when external pressures increase and risks are disproportionately shared by women and foreigners.

Social democratic welfare state societies favor the individual as the unit and agent of the life course. Their still-high degrees of social protection support the continuity across life, and this tends to standardize life courses. The income distribution is still quite equitable, and transfer incomes stabilize and equalize income trajectories. In the southern European familistic welfare societies the bulk of increasing risks of unemployment and of an aging population is loaded onto the families. In such a social context, the individual life course depends highly on the relative ability of families to cushion risks. This should increase cross-sectional and life course inequalities. Thus, considerable evidence has accumulated to demonstrate not only how patterns of life courses differ between countries of even similar levels of economic and social development but also how these differences are linked to specific institutional configurations and public policies.

MACRO-SOCIAL CONTEXT AND LIFE COURSE BEHAVIOR AND PROCESSES: THE MISSING LINKS

The reader is probably convinced by now not only that history, society, and institutions have large impacts on individual lives but also that it is possible to demonstrate in detail how life course outcomes and institutional configurations covary. However, students of human development traditionally want also to know how social contexts differentially affect individual development (and aging) and how, despite and in addition to social context, personal characteristics and processes forcefully influence life course outcomes. What is in the black box of the missing links between individual lives and macro-social contexts?

On a general level, one can differentiate six modes linking the macro-social context to characteristics and processes of persons (Diewald, 2000; Mayer, 2003). First, if one assumes that social contexts are highly variable, but that psychological dispositions as traits acquired in evolution are highly general and universal or normally distributed across populations, then knowing what is in the black box of cognitive, volitional, and emotional dynamics is interesting but adds little explanatory power. This is the basis both of using theoretical models of rational choice as "as–if" explanations (Coleman, 1990) that do not need to be ascertained empirically and of more modern versions of "nonrational" decision heuristics and experimental game theory (Gigerenzer & Todd, 1999). Second, because explicit selection processes are crucial components of gatekeeping in modern educational systems and work organizations, one must assume that social institutions differentially and effectively recruit members with psychological dispositions that they see as fitting to the tasks and environments they represent. The selection of workers either into public employment, with their corresponding bureaucratic mentalities, or into the private sector, are intuitive cases. Social actors will anticipate such selection criteria and might even adapt their self-representation in advance. A well-known case illustrating selection, self-selection, and anticipatory adaptation is the sorting of university students into different fields. For instance, students of business administration differ markedly in these regards from students of social work. Third, institutions socialize, and people accommodate to, the pressures of their social environments.

Fourth, types of social organization and modes of production "produce" corresponding personalities. Max Weber, for instance, in his famous study on farm labor in East Elbia, argued how the introduction of market forces into a feudal estate system favored the emergence of less deferential, less passive, more calculative, and more enterprising persons. Likewise, the entitlements and provisions of the welfare state that protect to some measure against the income losses resulting from illness, unemployment, maternity, and old age should favor the rise of other psychological dispositions than liberal market societies with much less protection would (Mayer, 1997; Mayer & Müller, 1986; see also Breen & Buchmann, 2002). For instance, many U.S. Americans constantly need to worry about restricted unemployment benefits, their health insurance coverage, the future yields of their retirement funds, and how their parents and themselves would be covered in case of chronic illness, whereas Swedes do not need to pay any attention to these matters. One would expect that U.S. Americans should show higher stress levels and stronger future orientations. Another good example is how different ways of institutionalizing the transition from school to the labor market require and reward different psychological resources and have consequences for psychological dispositions and personalities (Alfeld-Liro, Schnabel, Eccles, Baumert, & Barber, 2000). So far, all these suggested linkages suggest a primacy of the social contexts relative to the goals, values, and scope of actions of individual actors. However, it would be shortsighted—and indeed, misleading—to restrict oneself to a view of life courses where these would be seen exclusively or even predominantly as the outcome of social role playing, structural constraints, institutional regulation, and the socioeconomic circumstances of given historical periods. As much as life courses are the products of culture, society, and history, they are also the product of people as natural organisms, individual decision makers, and personalities (Diewald, 1999; Nollmann, 2003). Value orientations, for instance, play a large role in preferences for occupations over and above mere demand factors such as opportunities. The aftermath of the value changes in the 1970s (Inglehart, 1977) has provided a lot of prima facie evidence of how individuals not only are subject to their social contexts but also can actively and purposively influence and create them. Obvious instances are changes in living arrangements between men and women, preferred forms of child care, and so on.

There are also a priori reasons why an overly "sociologistic" construction of the life course is untenable. It would imply that actors are largely influenced by external factors (or chance). This would necessarily lead to beliefs in low efficacy of their own actions and would result in low self-esteem and low life satisfaction. Such a general outcome is inconsistent with the empirical evidence of psyches as positively self-equilibrating systems (e.g., in the Berlin Aging Study; Baltes & Mayer, 1999). The sociologistic position would also have to assume that, given the large inequalities in power, wealth, and prestige, most people would have to follow secondary, accommodative strategies rather than primary control strategies (Heckhausen, 1999; Mayer, 2003), which again does not seem to square with the empirical evidence.

One might well argue, therefore, that the genetic, physical, and psychological constraints on how people live out their lives and the interindividual variations resulting from them are not only non-negligible but also probably overwhelming compared with the determinants resulting from sociocultural differences (Rutter, 1997). It is obviously difficult, if not impossible, to assess such relative weights, although one might at least tender the hypothesis that across evolution, social and cultural construction and elaboration would tend to increase in their relative weight and other factors would recede in importance.

In stark contrast, however, Heckhausen (1999) argued why psychological modes of regulation of the life course should become more important than structural or institutional constraints. She made a distinction between *external* and *internal* regulation: External regulation is equated with social conditions such as legal sanctions, group pressure, or organizational rules; internal regulation is equated with relatively stable psychological dispositions related to modes of adaptation and coping or with regard to substantive preferences. Analogous to the theory of the civilizing process of Norbert Elias (1969), she claimed that "external enforcement via societal power has gradually, over centuries, been transformed into internalized rules and norms of conduct and behavior. This process of internalization renders the need for external societal enforcement obsolete" (Heckhausen, 1999, pp. 34–35). Heckhausen's position is consistent with sociologists' claim that life courses become deinstitutionalized; that cognitive biographical scripts about the normal life course become more important (Kohli, 1985); that institutions and traditional collectivities lose their binding power and that, therefore, individualization increases (Beck, 1986); and that more subtle forms of psychological influence have replaced the crude mechanisms of physical force and material incentives in controlling behavior (Foucault, 1977; Pizzorno, 1991).

Although one can hardly deny the historical thrust of the argument à la Elias and Foucault, I have many doubts as to its applicability to modern life courses as far as the role of normative orientations is concerned. According to John Meyer (1986), although internalized and strong religiously based norms in guiding life courses may have been appropriate and widespread in the 17th to 19th centuries, they would be highly dysfunctional in present-day societies, where very flexible situational adaptation is required (Mayer, 1988; Meyer, 1986). The relative importance that people and cultures accord to their lives as an overall developmental project is highly variable (Brandtstädter, 1990).

One rare instance in which one could empirically test which role psychological dispositions play in life course processes and whether they are responsive to changing social contexts is the transition of East Germany from a socialist society to a market economy. Indeed, psychologists have argued that personality characteristics should show most salience in times of sudden change and turbulence but are then also more vulnerable to modification (Caspi & Moffitt, 1993; Elder, 1974). In the context of a study of life courses during the transformation of East Germany after the fall of the Berlin Wall (Mayer et al., 1999), Diewald, Huinink, and Heckhausen (1996) examined first how control beliefs, control strategies, and feelings of self-respect varied among groups of different age and different occupational experiences before 1989 and between 1989 and 1993. Second, Diewald (2000) tested whether control beliefs had a net impact on unemployment, downward mobility, upward mobility, and occupational shifts between 1989 and 1993. It is noteworthy that control cognitions (as measured in 1993 and presumed to have been stable prior to this) played an important role in preventing unemployment but that they had no significant effects on upward and downward mobility. For the two variables already measured in 1991—internal control and fatalism—only fatalism showed any effect at all, and on only one of the four dependent variables: unemployment. In general, the evidence from these studies points more to psychological dispositions being the outcomes of (in this case, dramatic) life course events rather than being a strong influence on life course adaptation. At any rate, it is still a long way until the emerging differential life course sociology can be matched with a similarly differential study of the linkages between macro social contexts and human development.

My aim in this article was to illustrate and to elaborate how sociologists currently are constructing the life course analytically and how in their empirical research they are mapping varieties of life course regimes. Such a differential sociology of the life course has rapidly developed within the last 20 years, and it has augmented and progressed beyond the prior, more general conceptualizations of the social organization of human lives. Both detailed historical comparisons and cross-national research have contributed greatly not only in showing the variety of life course outcomes but also in suggesting close causal linkages to the web of institutions specific to given societies and periods. Along this scientific and intellectual journey, the different perspectives of psychology, demography, history, and sociology—that is, of human development, biography, aging, cohorts, and generations—have been refined and become distinct. Now is the time when a differential sociology of the life course needs to be reintegrated into the interdisciplinary study of human development, which in turn has to live up to the newly won knowledge about the force and variety of constraining and enabling social and historical contexts.

REFERENCES

Alfeld-Liro, C., Schnabel, K. U., Eccles, J., Baumert, J., & Barber, B. (2000, June). *Educational structure and development during the transition to adulthood in the USA and Germany.* Ms., 58.

Allmendinger, J. (1989a). *Career mobility dynamics: A comparative study of the United States, Norway and Germany* (Studien und Berichte Nr. 49). Berlin, Germany: Max-Planck-Institut für Bildungsforschung.
Allmendinger, J. (1989b). Educational systems and labor market outcomes. *European Sociological Review, 5,* 231–250.
Anderson, M. (1985). The emergence of the modern life cycle in Britain. *Social History, 10,* 69–87.
Aries, P. (1973). *Centuries of childhood.* Harmondsworth, England: Penguin.
Baltes, P. B., & Mayer, K. U. (Eds.). (1999). *The Berlin Aging Study: Aging from 70 to 100.* Cambridge, England: Cambridge University Press.
Beck, U. (1986). *Risikogesellschaft. Auf dem Weg in eine andere Moderne* [Risk society. On the road to another modernity]. Frankfurt am Main, Germany: Suhrkamp Verlag.
Bender, S., Konietzka, D., & Sopp, P. (2000). Diskonituität im Erwerbsverlauf und betrieblicher Kontext. [Discontinuity in working lives in the context of firms]. *Kölner Zeitschrift für Soziologie und Sozialpsychologie, 52*(3), 475–499.
Ben-Porath, Y. (1979). Family functions and structure and the organization of exchange. In International Union for the Scientific Study of Population (Ed.), *Economic and demographic change: Issues for the 1980's* (Vol. 3, pp. 51–64). Liège, Belgium: International Union for the Scientific Study of Population.
Bertaux, D. (Ed.). (1981). *Biography and society: The life history approach in the social sciences.* Beverly Hills, CA: Sage.
Blossfeld, H.-P., & Hakim, C. (Eds.). (1997). *Between equalization and marginalization: Women working part-time in Europe and the United States of America.* New York: Oxford University Press.
Boyer, R., & Durand, J.-P. (Eds.). (2001). *After Fordism.* Hampshire, NY: Palgrave.
Brandtstädter, J. (1990). Entwicklung im Lebensablauf. Ansätze und Probleme der Lebensspannen-Entwicklungspsychologie [Development in the life course. Approaches and problems of life span developmental psychology]. In K. U. Mayer (Ed.), *Lebensverläufe und sozialer Wandel* (Vol. 31, pp. 322–350). Opladen, Germany: Westdeutscher Verlag.
Breen, R., & Buchmann, M. (2002). Institutional variation and the position of young people: A comparative perspective. *Annals of the American Academy of Political and Social Science, 580,* 288–305. Thousand Oaks, CA: Sage.
Brückner, E., & Mayer, K. U. (1998). Collecting life history data: Experiences from the German Life History Study. In J. Z. Giele & G. H. Elder Jr. (Eds.), *Methods of life course research: Qualitative and quantitative approaches* (pp. 152–181). Thousand Oaks, CA: Sage.
Buchmann, M. (1989). *The script of life in modern society. Entry into adulthood in a changing world.* Chicago: University of Chicago Press.
Bude, H. (1995). *Das Altern einer Generation. Die Jahrgänge 1938 bis 1948* [The aging of a generation. The birth cohorts 1938 to 1948]. Frankfurt am Main, Germany: Suhrkamp Verlag.
Caspi, A., & Moffitt, T. E. (1993). When do individual differences matter? A paradoxical theory of personality coherence. *Psychological Inquiry, 4*(4), 247–271.
Clausen, J. A. (1986). *The life course: A sociological perspective.* Englewood Cliffs, NJ: Prentice Hall.
Coleman, J. S. (1990). *Foundations of social theory.* Cambridge, MA: Belknap.
Diewald, M. (1999). *Entwertungen, Umwertungen, Aufwertungen. Ostdeutsche Erwerbsverläufe zwischen Kontinuität und Wandel* [East German working lives after 1989: Between continuity and change]. Habilitation thesis for sociology, Faculty of Political and Social Sciences, Free University Berlin, Berlin, Germany.
Diewald, M. (2000). Continuities and breaks in occupational careers and subjective control: The case of the East German transformation. In J. B. Silbereisen & R. K. Silbereisen (Eds.), *Adversity and challenge in life in the new Germany and in England* (pp. 239–267). New York: Macmillan.

Diewald, M., Huinink, J., & Heckhausen, J. (1996). Lebensverläufe und Persönlichkeitsentwicklung im gesellschaftlichen Umbruch: Kohortenschicksale und Kontrollverhalten in Ostdeutschland nach der Wende [Life courses and developmental control in times of a macrosocial rupture: The case of different birth cohorts in the East German transformation process]. *Kölner Zeitschrift für Soziologie und Sozialpsychologie, 48,* 219–248.

DiPrete, T. A. (2002). Life course risks, mobility regimes, and mobility consequences: A comparison of Sweden, Germany, and the United States. *American Journal of Sociology, 108,* 267–309.

Eisenstadt, S. N. (1964). *From generation to generation: Age groups and social structure.* New York: Free Press of Glencoe.

Elder, G. H., Jr. (1974). *Children of the Great Depression.* Chicago: University of Chicago Press.

Elder, G. H., Jr. (2001). Life course: Social aspects. In N. Smelser & P. B. Baltes (Eds.), *International encyclopedia of the social and behavioral sciences* (Vol. 13, pp. 8817–8821). Cambridge, England: Elsevier.

Elias, N. (1969). *Über den Prozeß der Zivilisation: Soziogenetische und psychogenetische Untersuchungen* [The process of civilization: Sociogenetic and psychogenetic studies]. Bern, Switzerland: Francke Verlag.

Erikson, E. H. (1980). *Identity and the life cycle.* New York: Norton.

Esping-Andersen, G. (1990). *The three worlds of welfare capitalism.* Princeton, NJ: Princeton University Press.

Esping-Andersen, G. (1999). *Social foundations of postindustrial economies.* Oxford, England: Oxford University Press.

Foucault, M. (1977). *Discipline and punish: The birth of the prison.* New York: Pantheon.

Giddens, A. (1984). *The constitution of society: Outline of the theory of structuration.* Berkeley: University of California Press.

Gigerenzer, G., & Todd, P. M. (1999). *Simple heuristics that make us smart.* Oxford, England: Oxford University Press.

Grundmann, M. (1992). *Familienstruktur und Lebensverlauf. Historische und gesellschaftliche Bedingungen individueller Entwicklung* [Family structure and life course. Historical and social conditions of individual development]. Frankfurt am Main, Germany: Campus Verlag.

Hall, P. A., & Soskice, D. W. (Eds.). (2001). *Varieties of capitalism: The institutional foundations of comparative advantage.* Oxford, England: Oxford University Press.

Hareven, T. (1986). Historical changes in the social construction of the life course. *Human Development, 29,* 171–180.

Hareven, T. (Ed.). (1996). *Aging and generational relations: Life course and cross-cultural perspectives.* New York: Aldine de Gruyter.

Heckhausen, J. (1999). *Developmental regulation in adulthood: Age-normative and sociostructural constraints as adaptive challenges.* New York: Cambridge University Press.

Henz, U. (1996). *Intergenerationale Mobilität. Methodische und empirische Untersuchungen* [Intergenerational mobility. Methodological and empirical analyses]. Berlin, Germany: Max-Planck-Institut für Bildungsforschung.

Hernes, G. (1972). The process of entry into first marriage. *American Sociological Review, 37,* 513–547.

Hernes, G. (1976). Structural change in social processes. *American Journal of Sociology, 82,* 513–547.

Hillmert, S. (2001a). *Ausbildungssysteme und Arbeitsmarkt. Lebensverläufe in Grossbritannien und Deutschland im Kohortenvergleich* [Education, training systems and the labor market. A cohort comparison of life courses in Great Britain and Germany]. Wiesbaden, Germany: Westdeutscher Verlag.

Hillmert, S. (2001b). *Kohortendynamik und Konkurrenz an den zwei Schwellen des dualen Ausbildungssystems* [Cohorts and competition: Transitions from school to work in Germany in the context of economic and demographic change]. Working paper 2/2001 of the project "Education, Training, and Occuptation: Life Courses of the 1964 and 1971 Birth Cohorts in West Germany. Berlin, Germany: Max-Planck-Institut für Bildungsforschung.

Hillmert, S. (2002). Labour market integration and institutions: An Anglo-German comparison. *Work, Employment and Society, 16,* 675–701.

Huinink, J. (1995). *Warum noch Familie? Zur Attraktivität von Partnerschaft und Elternschaft in unserer Gesellschaft* [Do we still need the family? The attraction of partnerships and parenthood in our society]. Frankfurt am Main, Germany: Campus Verlag.

Huinink, J., Mayer, K. U., Diewald, M., Solga, H., Sorensen, A., & Trappe, H. (1995). *Kollektiv und Eigensinn. Lebensverläufe in der DDR und danach* [Collectivities and individuals. Life courses in the GDR and after the fall of the wall]. Berlin, Germany: Akademie Verlag.

Inglehart, R. (1977). *The silent revolution: Changing values and political styles among Western publics.* Princeton, NJ: Princeton University Press.

Kertesz, I. (1992). *Fateless.* Evanston, IL: Northwestern University Press.

Kohli, M. (1981). Biography: Account, text, method. In D. Bertaux (Ed.), *Biography and society: The life history approach in the social sciences* (pp. 61–75). Beverly Hills, CA: Sage.

Kohli, M. (1985). Die Institutionalisierung des Lebenslaufs. Historische Befunde und theoretische Argumente [The institutionalization of the life course. Historical evidence and theoretical arguments]. *Kölner Zeitschrift für Soziologie und Sozialpsychologie, 37,* 1–29.

Konietzka, D. (1999). *Ausbildung und Beruf. Die Geburtsjahrgänge 1919–1961 auf dem Weg von der Schule in das Erwerbsleben* [Education and occupation. The birth years 1919–1961 in the transition from school to work]. Opladen, Germany: Westdeutscher Verlag.

Krappmann, L. (2001). Die Sozialwelt der Kinder und ihre Moralentwicklung [The social world of children and their moral development]. In W. Edelstein, F. Oser, & P. Schuster (Ed.), *Moralische Erziehung in der Schule* (pp. 155–174). Basel, Switzerland: Beltz.

Kreppner, K. (1999). Beziehung und Entwicklung in der Familie, Konitinuität und Diskontinuität bei der Konstruktion von Erfahrungswelten [Relationships and development in the family. Continuity and discontinuity in the construction of reality]. In M. Grundmann (Ed.), *Konstruktivistische Sozialisationsforschung. Lebensweltliche Erfahrungskontexte, individuelle Handlungskompetenzen und die Konstruktion sozialer Strukturen* (pp. 180–207). Opladen, Germany: Suhrkamp.

Leisering, L. (2003). Government and the life course. In J. T. Martimer & M. J. Shanahan (Eds.), *Handbook of the life course* (pp. 205–225). New York: Kluwer Academic/Plenum.

Leisering, L., & Leibfried, S. (1999). *Time and poverty in the welfare state: United Germany in perspective.* Cambridge, England: Cambridge University Press.

Leisering, L., & Leibfried, S. (2001). Paths out of poverty: Perspectives on active policy. In A. Giddens (Ed.), *The global third way debate* (pp. 199–209). Cambridge, England: Polity Press.

Lerner, R. M., & Busch-Rossnagel, N. A. (Eds.). (1981). *Individuals as producers of their development: A life span perspective.* New York: Academic Press.

Levi, P. (1995). *Survival in Auschwitz.* New York: Touchstone Books.

Linton, R. (1945). *The cultural background of personality.* New York: D. Appleton-Century.

Mannheim, K. (1952). The sociological problem of generations. In P. Kecskemeti (Ed.), *Essays on the sociology of knowledge* (pp. 276–322). New York: Routledge & Kegan Paul.

Mayer, K. U. (1986). Structural constraints on the life course. *Human Development, 29,* 163–170.

Mayer, K. U. (1988). German survivors of World War II: The impact on the life course of the collective experience of birth cohorts. In M. W. Riley (Ed.), *Social structures and human lives: Social change and the life course* (Vol. 1, pp. 229–246). Newbury Park, CA: Sage.

Mayer, K. U. (1995). Gesellschaftlicher Wandel, Kohortenungleichheit und Lebensverläufe [Social change, cohort inequality and life courses]. In P. A. Berger & P. Sopp (Eds.), *Sozialstruktur und Lebenslauf* (pp. 27–47). Opladen, Germany: Leske + Budrich.

Mayer, K. U. (1997). Notes on a comparative political economy of life courses. *Comparative Social Research, 16,* 203–226.

Mayer, K. U. (2001). The paradox of global social change and national path dependencies: Life course patterns in advanced societies. In A. E. Woodward & M. Kohli (Eds.), *Inclusions and exclusions in European societies* (pp. 89–110). London: Routledge.

Mayer, K. U. (2003). The sociology of the life course and lifespan psychology: Diverging or converging pathways? In U. M. Staudinger & U. Lindenberger (Eds.), *Understanding human development: Dialogues with lifespan psychology* (pp. 463–481). Boston: Kluwer Academic.

Mayer, K. U., & Carroll, G. R. (1990). Jobs and classes: Structural constraints on career mobility. In K. U. Mayer & N. B. Tuma (Eds.), *Event history analysis in life course research* (pp. 23–52). Madison: University of Wisconsin Press.

Mayer, K. U., Diewald, M., & Solga, H. (1999). Transitions to post-Communism in East Germany: Worklife mobility of women and men between 1989 and 1993. *Acta Sociologica, 42,* 35–53.

Mayer, K. U., & Hillmert, S. (2003). New ways of life or old rigidities? Changes in social structures and life courses and their political impacts. In H. Kitschelt & W. Streeck (Eds.), *West European politics* (Special issue on Germany: Beyond the Stable State, Vol. 26, No. 4, pp. 79–100). London: Frank Cass. [Also published 2004 in H. Kitschelt & W. Streeck (Eds.), *Germany: Beyond the stable state* (pp. 79–100). London: Frank Cass.]

Mayer, K. U., & Huinink, J. (1990). Age, period, and cohort in the study of the life course: A comparison of classical A-P-C analysis with event history analysis or farewell to LEXIS? In D. Magnusson & L. R. Bergman (Eds.), *Data quality in longitudinal research* (pp. 211–232). Cambridge, England: Cambridge University Press.

Mayer, K. U., & Huinink, J. (1993). Lebensverläufe und gesellschaftlicher Wandel: Von der Kohortenanalyse zur Lebensverlaufsanalyse [Life courses and social change: From cohort analysis to the study of life courses]. In R. Hauser, U. Hochmuth, & J. Schwarze (Eds.), *Mikroanalytische Grundlagen der Gesellschaftspolitik* (pp. 92–111). Berlin, Germany: Akademie Verlag.

Mayer, K. U., & Müller, W. (1986). The state and the structure of the life course. In A. B. Sorensen, F. E. Weinert, & L. R. Sherrod (Eds.), *Human development and the life course: Multidisciplinary perspectives* (pp. 217–245). Hillsdale, NJ: Lawrence Erlbaum Associates, Inc.

Mayer, K. U., & Schoepflin, U. (1989). The state and the life course. *Annual Review of Sociology, 15,* 187–209.

Meyer, J. W. (1986). The self and the life course: Institutionalization and its effects. In A. B. Sorensen, F. E. Weinert, & L. R. Sherrod (Eds.), *Human development and the life course: Multidisciplinary perspectives* (pp. 199–216). Hillsdale, NJ: Lawrence Erlbaum Associates, Inc.

Mills, M., & Blossfeld, H.-P. (2001). A causal approach to interrelated family events: A cross-national comparison of cohabitation, nonmarital conception, and marriage. *Canadian Studies in Population, 28,* 409–437.

Modell, J. (1991). *Into one's own: From youth to adulthood in the United States 1920–1975.* Berkeley: University of California Press.

Modell, J., Furstenberg, F. F., Jr., & Hershberg, T. (1976). Social change and transition to adulthood in historical perspective. *Journal of Family History, 1*(1, Autumn), 7–32.

Myles, J. (1990). States, labor markets and life cycles. In R. Friedland & A. F. Robertson (Eds.), *Beyond the marketplace: Rethinking economy and society* (pp. 271–298). New York: Aldine de Gruyter.

Myles, J. (1993). Is there a post-Fordist life course? In W. R. Heinz (Ed.), *Institutions and gatekeeping in the life course* (pp. 171–185). Weinheim, Germany: Deutscher Studien-Verlag.

Nollmann, G. (2003). Warum fällt der Apfel nicht weit vom Stamm? Die Messung subjektiver intergenerationaler Mobilität [Like father, like son. The measurement of subjective intergenerational mobility]. *Zeitschrift für Soziologie, 32,* 123–138.

Oppenheimer, V. K., & Kalmijn, M. (1995). Life cycle jobs. *Research in Social Stratification and Mobility, 14,* 1–38.

Parsons, T. (1942). Age and sex in the social structure of the United States. *American Sociological Review, 7,* 604–616.
Piaget, J. (1970). *Structuralism.* New York: Basic Books.
Pizzorno, A. (1991). Social control and the organization of the self (a summary of the original paper). In P. Bourdieu & J. S. Coleman (Eds.), *Social theory for a changing society* (pp. 232–234). New York: Russell Sage Foundation.
Riley, M. W., Kahn, R. L., & Foner, A. (Eds.). (1994). *Age and structural lag: Society's failure to provide meaningful opportunities in work, family, and leisure.* New York: Wiley.
Rowntree, B. S. (1914). *Poverty: A study in town life.* London: Nelson. (Original work published 1901)
Rutter, M. (1997). Nature–nurture integration: The example of antisocial behavior. *American Psychologist, 52,* 390–398.
Ryder, N. B. (1965). The cohort as a concept in the study of social change. *American Sociological Review, 30,* 843–861.
Ryder, N. B. (1980). *The cohort approach: Essays in the measurement of temporal variations in demographic behavior.* New York: Arno.
Sennett, R. (2000). *The corrosion of character.* New York: Norton.
Solga, H. (2003). Das Paradox der integrierten Ausgrenzung von gering qualifizierten Jugendlichen [The paradox of integration and exclusion of less-educated youth]. *Aus Politik und Zeitgeschichte, B 21–22,* 19–25.
Sørensen, A. B. (1990). Processes of allocation to open and closed positions in social structure. In J. B. Zelditch & M. Zelditch Jr. (Eds.), *Sociological theories in progress* (Vol. 3, pp. 256–287). New York: Sage.
Thomas, W. I., & Znaniecki, F. (1918). *The Polish peasant in Europe and in America: Monograph of an immigrant group.* Chicago: University of Chicago Press.
Thompson, E. P. (1976). *The making of the English working class.* Harmondsworth, England: Penguin.
Winter, J. (1986). *The Great War and the British people.* Cambridge, MA: Harvard University Press.

Civic Engagement, Political Identity, and Generation in Developmental Context

Abigail J. Stewart and Christa McDermott
University of Michigan

We provide a developmental account of civic engagement—specifically, political participation. Civic engagement is more likely among people with politicized identities, an activist stance, and an interest in diverse peers. The form of civic engagement (focused on transmission of parental values or on social change) is shaped by different generations' relative tendency to identify horizontally (with each other) or vertically (with previous generations). Adolescence is proposed as a formative period for the development of a politicized identity and student activism as providing an opportunity to develop both personal efficacy in the political realm and experience working toward a goal with diverse peers. The intersection of late adolescence with periods of intense social discontinuity increases within-generation identification and decreases interest in cross-generational transmission of values. Young American women in the middle to late 20th century experienced such a confluence of factors, and we focus on studies of women's political participation and development of politicized identities at this time. Thus, forms of civic engagement are shaped not only by individual experiences but also by cohort or generational identity.

A few years ago, political scientist Robert Putnam (1995, 2000) asserted that Americans are "bowling alone" more now than in an earlier era. He argued that, unlike previous generations, we have severed our ties to our communities by allowing our traditionally lively "voluntary associations" to wither away, leaving a relatively depopulated space where there used to be lively "civil society" that was distinct from both government and the marketplace.

Putnam's (1995, 2000) argument has been countered from different angles, including two that are pertinent for us (see, e.g., McLean, Schultz, & Steger, 2002). The first counterargument emphasizes the inevitability of historical change in the

---

Requests for reprints should be sent to Abigail J. Stewart and Christa McDermott, 204 State Street, Lane Hall, Ann Arbor, MI 48109–1109. E-mail: abbystew@umich.edu and mcdc@umich.edu

nature, purposes, and forms of voluntary association or civic engagement. Here scholars argue that although Putnam is correct that certain forms of voluntary association (e.g., men's service clubs) declined between the 1950s and the 1990s, they have been replaced by other forms (Rosenblum, 1998; Wuthnow, 1998). These scholars advise that it is critical to cast a broad net when attempting to assess individuals' methods of "connecting" or "associating" outside of both government and economic relations (see Burns, Schlozman, & Verba, 2001). Some of these scholars point as well to the fact that the political significance of these different forms of connecting and associating alter depending on the context and the purposes the individual brings to the group.

The second counterargument builds on the first, arguing not only that forms of civic engagement change over time but also that the changes may be related to generational experiences. Thus, particular generations, within a particular society, may be exposed to one set of models for civic engagement and may create a particular set of associations that serve the needs of that generation. We may expect, then, both *historical change* in patterns of civic engagement and *generation* to play some role in structuring the pattern within a single period, or for different generations at the same time but at developmentally different life stages.

Finally, some social theorists have noted that the meanings and consequences of civic engagement vary by the person. Rosenblum (1998) argued

> that voluntary associations are personally and politically evocative and changing, and ... the vicissitudes of personal life affect what we take from membership. The significance of association depends on the experiences individuals bring to it, including the unique obstacles we each face in cultivating and exhibiting particular competencies and dispositions. (p. 8)

She made clear that civic engagement should have a *developmental course*; an individual's experiences are brought to that engagement and shape its impact, and the acts associated with civic engagement, in turn, have consequences in shaping the person. Rosenblum's argument suggests that developmental psychologists have something to contribute to the larger discussion of the generational and developmental features of civic engagement. In this article we focus on only one feature of the wide array of activities that connect the individual to a larger society and therefore are defined as "civic engagement": political participation.

There are many ways for individuals to be involved in their community, including micro relations of face-to-face contact with other individuals, extended family networks, participation in community associations, and large-scale political movements. All too often, pundits and researchers lump these forms of civic engagement together. This is problematic because it collapses into a single index the individual's relationships with neighbors and friends, a range of community institutions, and larger social institutions, including the government. In fact, Putnam (2000) broke civic engagement into three categories: (a) church, (b) work, and (c) community. *Community-oriented engagement* includes activities as diverse as political participation, on a scale ranging from voting to running for office, as well as his celebrated bowling leagues, PTAs, and the Red Cross (Putnam, 2000). He also tracks civic engagement almost exclusively through the changing numbers of memberships with formal groups, primarily politically moderate groups. This means Putnam pays little attention to many forms of civic involvement, especially the very local (e.g., tenants' associations), informal (self-help groups), and the more socially and politically extreme groups (gangs, Promise Keepers, Greenpeace) that have grown since the 1960s (Boggs, 2002). Moreover, one of the features of the landscape that Putnam views with concern (withdrawal of trust from the government) may actually have been fueled both by social movements of the late 1960s and the conservative groups that emerged in response to them. If so, then this new, polarized set of civic engagements may in a sense lie behind a phenomenon Putnam views as evidence of disengagement (Snyder, 2002). Therefore, as we focus on political forms of civic engagement, we include different types of participation that may vary in their appeal to different sensibilities and at different life stages, particularly protest politics (most attractive in late adolescence) and participation in mainstream formal politics (more attractive in middle age) and specifically women's engagement in each.

It may seem odd to focus on political participation. We know that relatively few American adults actually do participate in politics at any point in their adult lives, and it is arguable that women are less likely to do so than men (see Burns et al., 2001). We believe, though, that political participation is not only an important feature of social life, but it is also a feature that psychologists can help us understand. The women's movement of the late 1960s and early 1970s certainly encouraged more political participation among women in the United States. Many living American adult women were influenced by that movement and may be expected to be unusually politically active (see Cole, Zucker, & Duncan, 2001; Cole, Zucker, & Ostrove, 1998; Sigel, 1996). In addition, there is always a burst of interest among women in education, labor force participation, and politics, even in feminism, when the demographic situation involves a sex ratio according to which there are "too many women" or "too few men" (Guttentag & Secord, 1982; e.g., in England after World War I, when many men died, and in the United States during World War II, when the economy depended on women's work in the labor force). Moreover, there are always some people, including some women, who get engaged in the larger political world and feel that they have a role to play in it, regardless of their generation or the moment in history. There is, then, also a role for disposition or personality in shaping political participation (Sullivan & Transue, 1999).

Under what circumstances do women (who traditionally are viewed in the United States as relatively less prone to engage with the political domain) see themselves as involved in the political arena? When do they actually form personal identities around their political views and get involved in sustained political participation? To what extent does generation play a role in defining the terms of women's "civic engagement"? In answering these questions, we draw on empirical research to organize a developmental account of civic engagement in the political sphere that might help us understand other aspects of civic engagement as well. This developmental account focuses on adolescence as a formative period for creating a lifetime of civic engagement through three avenues: (a) developing a politicized identity, (b) adopting activism as a stance toward the political sphere, and (c) developing an interest in diverse peers. We consider evidence that this formative period both shapes individual development and creates distinctive generations or cohorts that have average tendencies toward cross-generational versus peer identification, tendencies that may have consequences for the civic engagement of generation members (see Rogler, 2002; Settersten, 1999, for related arguments).

FORMATION OF POLITICAL IDENTITIES

First, we consider evidence that adolescence is an important formative period for developing a politicized identity. In the development of political identity and political action, certain kinds of events are significant for many groups or individuals. In fact, Mannheim (1928/1972) pointed to this phenomenon when he noted the "problem of generations." He observed that individuals who experience the same important social events during their formative years tend to share a generational stance toward those events that differs from that of members of other generations who experience these same events at other developmental stages. In U.S. history we have particularly strongly identified generations associated with the Depression and World War II, and (variously) the "baby boom" and "the 1960s." There are also important subgroups of the population who identify themselves with particular events. For example, U.S. women of different generations may identify themselves with obtaining the right to vote, with the women's movement, with the *Roe v. Wade* legal-access-to-abortion decision, or with Title IX's assurance of equality of access to athletics in the educational context. Within the baby boom generation, then, some individuals identify with the women's movement generally, and some identify only with specific political events within the movement.

Similarly, African Americans of different generations may define themselves in terms of the civil rights movement and the striking down of "Jim Crow" practices or with affirmative action. Among African American baby boomers, some may identify with the civil rights movement generally, some with "Black Power," and others with affirmative action. Still others may have experienced other events (e.g., the Vietnam war, the women's movement) as formative, and some may have experienced none of these events as formative. These *generation units* (according to Mannheim, 1928/1972) are groups defined within a single generation, with a common experience of social history and social identity. They "are characterized by . . . an identity of responses, a certain affinity in the way in which all move with and are formed by their common experience" (Mannheim, 1928/1972, p. 122).

At the individual level, a number of factors relate to the development of a politicized identity (see Cole, Zucker, & Duncan, 2001; Stewart & Healy, 1986). These include coming from a political family (e.g., "red diaper babies," who were the children of leftists), personal experiences of discrimination or privilege that crystallize an individual's understanding of or commitment to an ideology, formal education or exposure to an ideology (e.g., exposure to women's studies or feminist theory), and personal dispositions (e.g., cognitive and personality styles that make it more likely to see oneself on the political stage).

The most powerful overall impact may occur, though, when certain individuals (e.g., those raised in political families, with certain dispositions, etc.) are exposed to powerful social events during their identity-formative years (see Jennings, 2002; Stewart & Healy, 1989). This intersection of the social and the individual seems particularly likely to catalyze political identification, and it arises most often when the social experiences associated with an individual's adolescence are strikingly discontinuous from those experienced in childhood. In this case, the salience of the social experience is particularly intense and therefore formative. A series of studies have examined whether individuals from different generations experience the same social events differently, as well as whether events experienced in adolescence tend to be particularly formative of political consciousness (see also Delli Carpini, 1989; Schuman & Scott, 1989).

THE MEANINGFULNESS OF EVENTS: LIFE STAGE AND GENERATION-BASED DIFFERENCES

Stewart and Healy (1989) disaggregated Eli Ginzberg's (1966) and Alice Yohalem's (1979, 1993) sample of female graduate students at Columbia University in the postwar period into three cohorts and examined the impact of World War II on those three cohorts. They found that the impact was strongest—both as subjectively reported and in terms of the complex impact it had on later experience of labor force participation—for women who were in the identity-formative late adolescent/young adulthood stage during World War II rather than the Depression or the postwar period.

Duncan and Agronick (1995) examined college-educated women from two different selective private women's colleges, drawn from different cohorts, to conduct the same kind of tests. They found that for four out of five samples, social events coinciding with the same life stage (late adolescence/early adulthood) were viewed as more personally meaningful than events drawn from other periods. Because the samples were drawn from three different age cohorts, this study showed that different events were salient and meaningful (identity formative) even for women drawn from cohorts relatively close in age and that the same events might be salient and meaningful (identity formative) for women from one cohort but not for those from another (who experienced that event later or earlier in their own development). The nonsignificant results for the fifth sample nevertheless showed the same overall pattern in distribution of meaningful events by life stage.

Duncan and Agronick's (1995) results were particularly impressive because they did not have access to identical measures in the five different samples; instead, they were forced to code open-ended responses to slightly different questions in order to assess the meaningfulness of events. Zucker (1998) used identical measures to examine women who graduated from the University of Michigan in the early 1950s, 1972, and 1992. Because she asked all the women about the same events drawn from three different periods (pre-1960s, 1960s era, and post-1960s), she was able to make three points: (a) events from a given period were rated as significantly more meaningful by the cohort that was in late adolescence/early adulthood at the time of those events, and thus all three cohorts attached the most meaning to the events they experienced in late adolescence/early adulthood; (b) the same events were rated differentially by the three cohorts (although 1960s events, especially the women's movement, were rated as relatively meaningful by all three cohorts); and (c) the tendency to differentiate among events drawn from the three periods was strongest for the cohort that came of age in the 1970s. (On a 3-point scale, the average rating difference between the highest rated set of events and the next highest rated set was .61 for that cohort, whereas it was only .02 for the youngest cohort and .05 for the oldest; see Table 10 in Zucker, 1998.) In short, social events experienced in late adolescence were formative for all three groups but, as Duncan and Agronick thought they had shown with somewhat weaker data, those events played a particularly strong role for the cohort that came of age in the late 1960s and early 1970s.

This tendency to differentiate sharply between formative events and other social events might characterize cohorts or generations that are particularly *cohort identified*, or disposed toward horizontal (rather than vertical or cross-generational) relations. In a study of the graduates of a midwestern high school that included both men and women, Stewart (2003) used the same kind of data to assess two issues: (a) to which events each cohort attached personal meaning and (b) whether respondents who graduated in the mid-1950s were less likely than those who graduated in 1968 to differentiate sharply between events drawn from different periods in social history. The changing racial composition of the school, and the increasingly active local politics regarding school integration, offered a unique opportunity to study the development of political identity in young adults of similar backgrounds but who lived in very different sociopolitical settings. In the mid-1950s, Midwest High School (the name of the high school has been changed) was 25% African American. By 1968, African Americans constituted 50% of the school population. The school was closed that year as part of a district desegregation plan. At the same time, local and national politics surrounding civil rights, school integration, and other social movements had greater prominence in the students' lives than it did in the 1950s cohort. Thus, the two cohorts studied were separated by only a decade but experienced very different high school environments and political atmospheres. Like the previous studies, in this study Stewart found that each cohort scored significantly higher than the other on attaching personal meaning to the events drawn from the period in which they themselves came of age. In addition, she also found that the 1960s cohort made a much sharper distinction in their attachment of personal meaning to events between events drawn from the period of their coming of age and that of their parents or the comparison cohort (see Figure 1). This political identification with their own cohort means they felt most influenced by the social events of their young adulthood and shared with their peers the values emerging from the social upheaval. In contrast, the 1950s graduates identified equally with the formative events of their parents, their own adolescence, and those of the period after.

Why might these two cohorts differ in this way? One possibility is that students who came of age in the late 1960s and early 1970s were in fact more likely than those who came of age before or after to have actually participated in student protest politics during those years. In short, it might be that these individuals were affected not only by the coincidence of their developmental stage and the historical moment but also by their own actions during that stage. Such a difference in early experience may be especially important in examining women's political participation.

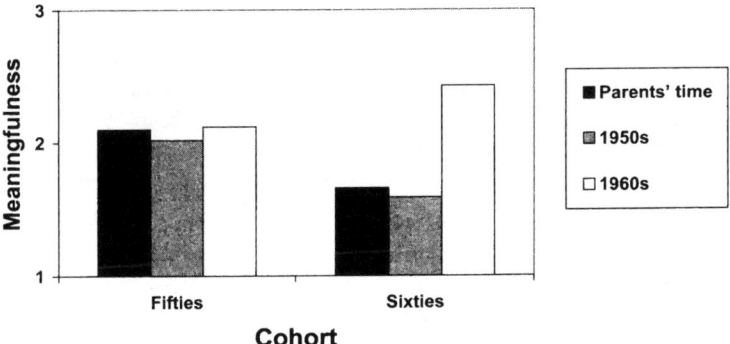

FIGURE 1 Personal meaningfulness of historic events rated by midwest high school graduates. Reprinted from "Gender, Race and Generation in a Midwest High School: Using Ethnographically Informed Methods in Psychology" by A. J. Stewart. *Psychology of Women Quarterly, 27*, 1–11. Published by Blackwell. Copyright © 2003. Used by permission.

STUDENT ACTIVISM AND CIVIC ENGAGEMENT

A number of scholars have studied the long-term impact of student activism on activists, showing that in general individuals who make a major commitment to activism as young adults remain unusually politically engaged in later adulthood (Braungart & Braungart, 1990; DeMartini, 1983; Fendrich, 1993; Jennings, 2002; McAdam, 1988). These studies have only occasionally focused on women (see McAdam, 1992, for an example), and have generally been limited to students at the extreme end of political participation. Cole and Stewart (1996) studied both African American and European American women who graduated from the University of Michigan to explore the role of student activism in the creation of political identity in a more "ordinary" sample of college students, some of whom were active and some not. One sample was part of a longitudinal study initiated by Tangri of a stratified random sample of the class of 1967, which was virtually all White (see Tangri & Jenkins, 1993, for an overview of the study). The second was a sample of African American women who graduated between 1967 and 1973, a period during which admission of African American students increased fairly rapidly. Both of these samples attended one of the most politically activist college campuses in the United States at a time of intense activism, but of course only a minority of students in fact participated in political protest. Thus, these samples offered an opportunity to examine the long-term impact of adolescent political engagement, in the form of student activism.

Among the European American women, Cole and Stewart (1996) found that student activism predicted midlife political consciousness, which in turn predicted midlife political participation. Among the African American women, both student activism and political consciousness predicted midlife political participation, but activism did not predict political consciousness. Although the data from the European American women were partly longitudinal, the African American sample was cross-sectional; in both cases, midlife political consciousness and participation were measured concurrently. However, the likeliest explanation for the different findings lies not in these methodological issues but in the fact that the African American women were higher than the European American women in all three indicators of politicization. Thus, they may simply have been a more politicized sample in every sense, and therefore for them student activism may not have played the same kind of decisive formative role that it did for the European American women.

In a related study, Cole et al. (1998) found that European American student activists at the University of Michigan from the same period (late 1960s and early 1970s) were higher in midlife political efficacy, political salience, and collectivist orientation, as well as political participation, than their nonactivist counterparts. Cole et al. (1998) concluded that this early political participation may have served to initiate this group of White women into a lifetime of political engagement. In an analysis of data from both the European American longitudinal and the African American concurrent samples, Stewart, Settles, and Winter (1998) explored two different forms of involvement with these movements. They separated activists from *engaged observers* (people who followed, and gave moral support to, the social movements of the time, but were not activists), and from a third group of nonparticipants (i.e., women who reported that the movements were not meaningful to them and that they did not engage in activism in them). In these analyses, Stewart et al. looked at the women's movement and the civil rights movement separately for the two groups of women. First, both movements produced high levels of participation and engagement in both samples, although European American women were less involved with the civil rights movement than the African American women were. In fact, there were so few African American women who were not involved that they could really compare self-reported activists only to women who were either engaged observers or nonparticipants. For both movements, activists and engaged observers reported that the movement had a strong impact on them, and the nonparticipants did not. Thus, simply being "engaged" by a movement in adolescence (but not active in it) resulted in midlife reports that the movement had personal meaning in and impact on a woman's life. However, women who had been activists as students were significantly more likely than both of the other groups (who did not differ from each other) to report high levels of midlife political participation. Stewart et al. concluded that the experience of student activism may have enhanced the women's sense of political efficacy or agency in a crucial way that enabled that group to continue to take political actions, even in midlife. In addition to enhancing agency, it is of course also possible that an important feature of student activism is that it leads students to work closely with others who share their political goals and certain values but who may be very different in other ways. The experience of working hard with diverse peers for a valued end may be an important feature of activist experience that has consequences for later civic engagement, particularly engagements that target a person's larger community.

CONTACT WITH DIVERSE PEERS AND CIVIC ENGAGEMENT

Many studies have shown that student participation in community service activities and other kinds of school-based civic engagement predicts later adult engagement in the larger civic and political community. Hanks and Eckland (1978) used longitudinal data to examine the long-term effect of extracurricular school activity in high school and found that participation in those activities, among both women and men, was associated with participation in adult secondary associations at age 30, which in turn was associated with voting and low political alienation. Youniss, McLellan, and Yates (1997) reviewed the evidence that youth civic engagement in activities such as 4-H, Boy Scouts, YMCA, and other community organizations predicts later adult participation in comparable kinds of activity. They concluded that activity of this kind produces "civic identities" that mediate adult participation in building and maintaining communities. Although this seems plausible to us (and is parallel to our argument about politicization of identity), we suspect that additional features of youth activism may be important. We have already explored the importance of personal agency or efficacy as an important issue. We suspect that in addition to these factors, positive and successful collaboration with diverse peers may be a critical factor in increasing adult appreciation of differences among community members. In most communities, participation in civic and political activities depends on the capacity to work relatively closely, at least on limited goals and projects, with diverse community members—and, ideally, to enjoy it. Early competence in this kind of collaboration may in fact lead certain young people into student civic engagement, but participating in it is also likely to encourage the growth of this competence (see also Sullivan & Transue, 1999).

Using the same sample of graduates from a midwestern high school described earlier, Winter, Stewart, Henderson-King, Henderson-King, and Lewis (2003) examined the longitudinal impact of high school exposure to diversity on White adults' racial attitudes in middle age. Student participation in interracial extracurricular activity was assessed on the basis of yearbook lists and photographs. Adult racial attitudes were assessed in questionnaires administered more than 30 years after graduation. For White men, interracial activity in high school was in fact a strong positive predictor of positive racial attitudes in middle age. For women, interracial activity did not predict racial attitudes in the same way, perhaps because women's roles in extracurricular activities were so much less often (formal or informal) leadership roles at that time. Instead, having a politicized identity (i.e., attaching personal meaning to the women's movement, civil rights, and the Vietnam war) did predict positive racial attitudes for the women. Winter et al. proposed that for the men, incremental political socialization may have taken place in creating positive racial attitudes; in contrast, for the women, discontinuous or transformative political socialization (i.e., socialization associated with movements challenging the status quo) superseded any incremental effects. This may point to the particular importance of experiences of efficacy, or agency, in women's political socialization, even in the context of exposure to diverse peers.

GENERATIONAL IDENTITY AND CROSS-GENERATIONAL TRANSMISSION OF VALUES AND ATTITUDES

Stewart and Healy (1989), like Mannheim (1928/1972), argued that some generations have strong generational or cohort identities; in these generations, individuals adopt similar stances toward the social events that shaped them and feel a common bond or tie. It stands to reason that in emphasizing this common generational experience these generations must pay less attention to the other kinds of differences among them. At the same time, generationally identified cohorts must emphasize the differences between their generation and other generations, perhaps especially their parents' generation. This was certainly observed in the critical stances toward their elders of the famous English "Generation of 1914" and of the "baby boom" or Woodstock generation. It is interesting that although we have these theories that suggest that different cohorts will vary not only in generational identity but also in attitude toward parental generations, our theories of intergenerational transmission are generally universal; that is, they assume that all generations are equally likely (or unlikely) to seek to transmit or pass on values and knowledge from their parents' generation to their children's.

We suggest that it is in fact improbable that generations are equivalent in this respect. First, different generations are likely to find the values espoused by their parents' generations differentially meaningful to them. Indeed, Stewart (2003) found that the high school graduates of the 1950s were much more likely than high school graduates of the 1960s to find personal meaning in the events associated with their parents' coming of age (the Depression and World War II). To an equal degree, as mentioned earlier, the 1960s cohort made a much sharper distinction than the 1950s cohort between events of their own coming of age and all other events. Stewart outlined the differences in values in the two cohorts, emphasizing that the 1950s cohort adopted the core values of their parents, and saw themselves as trying to pass them on, whereas the 1960s cohort was much more likely to identify with social change in general, and changes associated with the 1960s in particular. The 1950s graduates developed political identities that were in-group oriented; their identities revolved around the values of a strong work ethic, preservation of family, and religious traditions their parents instilled in them. These values were often expressed in the form of civic involvement along the lines of what Putnam (1995, 2000) has tracked: voluntary associations such as Elks clubs; church groups; and card clubs born out of work, family, and religious networks. However, as forms of civic activity shifted in the 1960s to issues and groups beyond those based in work, family, and religion, and promoted social change directly affecting these values, the 1950s cohort's concept of civic-mindedness didn't translate. Committed to their parents' values, most were suspicious of social change and civic involvement that encouraged that change. It is important to note, then, that although this cohort has an attachment to the values of the same cohort (adults in the 1950s—their parents) that Putnam (1995, 2000) has tended to glorify as a "civic generation," the lesson learned by these now-adult children of that generation was very much an in-group–affirming message that did not lead to political engagement in adulthood. This fact underscores the complexity of defining, assessing, and charting changes over time in civic engagement.

It remains for future research to explore in detail how attention to the particularities of generation might affect our interpretation of findings about cross-generational transmission of (for example) political values and attitudes. One suggestive study is Jennings, Stoker, and Bowers's (1999) analysis of the cross-generational continuities in political attitudes between adult children who graduated from high school in 1965 and their parents on the one hand, and the 1965 graduates and their own children on the other. By looking at transmission across the two generations, Jennings et al. were able to assess whether transmission patterns for the oldest to the middle generation were similar to those for the middle to the youngest generation. Overall, they did find evidence for similar patterns in which aspects of political values get passed on, but they also found that the degree of politicization of the family was an important predictor of transmission as well. It is likely that families are especially politicized during precisely those periods when generationally defined cohorts are being created, and it is those cohorts that should be least inclined toward generational transmission of specific political attitudes. It is interesting that Jennings et al. found evidence that "adolescents emerging from highly politicized homes in 1965 were less likely to adopt the parental position on school integration than were adolescents from apolitical homes." They concluded that

> On the one hand, then, having a politicized family environment typically encourages the child to learn from the parent and to adopt the parent's views. On the other hand, it also leaves the child more attuned to outside political influences. In periods of upheaval like those of the mid-1960s, or in general when the political environment contains forces antithetical to parental inclination, this politicization may work against within-family congruence. Understanding how political engagement plays out in such cases, and tracing its implications for aggregate intergenerational change, constitutes another important challenge for future research. (pp. 23–24)

More studies are needed that compare transmission in different cohorts to assess how the family environment may interact with the larger cohort social environment in shaping different kinds of political engagement.

IMPLICATIONS OF FINDINGS

Our selective review of literature on political participation confirms our suggestion that an adequate framework for understanding individual women's relations with the political arena must include attention to developmental and generational issues. It seems clear that adolescence has a particularly important role both in the formation of politicized identities (and those that are not politicized) and in the creation or absence of a generational identity. Although we probably cannot and should not try to affect the kinds of social experiences whole generations encounter, we can offer them adequate information to permit them to care about the wider social world. Also, although the evidence is not conclusive about the reasons for the importance of political activism in adult politicization, it seems likely that personal efficacy or agency, and positive experiences with diverse peers, are both important features that deserve further attention. It is surely within our power to create more opportunities for young people to act meaningfully in concert with diverse peers on matters of community importance. Finally, it is clear that most of our cross-generational transmission studies focus on one cohort. As a result, it is impossible to be sure whether we can safely generalize the results of the Midwest High School study that suggest that generational transmission may vary as a function of the degree of generational identification of the cohort itself. If future research suggests that we can, then we may need to develop different strategies appropriate to encouraging political participation and civic engagement among young people in generations that differ in cohort identification and in openness to transmission of parental values.

ACKNOWLEDGMENTS

We are grateful to Jacquelynne Eccles and David G. Winter for thoughtful feedback on the first version of this manuscript and to Alyssa Zucker for assistance in making data available to us. We also benefited from valuable feedback from faculty and student members of the October 2003 LIFE Program's responses to a presentation of an earlier version of this article at the Max-Planck Institute in Berlin, Germany.

REFERENCES

Boggs, C. (2002). Social capital as political fantasy. In S. L. McLean, D. A. Schultz, & M. B. Steger (Eds.), *Social capital: Critical perspectives on community and "bowling alone"* (pp. 184–200). New York: New York University Press.

Braungart, M. M., & Braungart, R. G. (1990). The life-course development of left- and right-wing youth activist leaders from the 1960s. *Political Psychology, 11,* 243–282.

Burns, N., Schlozman, K. L., & Verba, S. (2001). *The private roots of public action.* Cambridge, MA: Harvard University Press.

Cole, E. R., & Stewart, A. J. (1996). Meanings of political participation among Black and White women: Political identity and social responsibility. *Journal of Personality and Social Psychology, 71,* 130–140.

Cole, E. R., Zucker, A. N., & Duncan, L. E. (2001). Changing society, changing women (and men). In R. K. Unger (Ed.), *Handbook of the psychology of women* (pp. 410–423). New York: Wiley.

Cole, E. R., Zucker, A. N., & Ostrove, J. M. (1998). Political participation and feminist consciousness among women activists of the 1960s. *Political Psychology, 19,* 349–371.

Delli Carpini, M. X. (1989). Age and history: Generations and sociopolitical change. In R. S. Sigel (Ed.), *Political learning in adulthood* (pp. 11–55). Chicago: University of Chicago Press.

DeMartini, J. (1983). Social movements participation. *Youth and Society, 15*, 195–223.
Duncan, L. E., & Agronick, G. S. (1995). The intersection of life stage and social events: Personality and life outcomes. *Journal of Personality and Social Psychology, 69*, 558–568.
Fendrich, J. (1993). *Ideal citizens*. Albany: State University of New York Press.
Ginzberg, E., and others. (1966). *Life styles of educated women*. New York: Columbia University Press.
Guttentag, M., & Secord, P. (1982). *Too many women? The sex ratio question*. Beverly Hills, CA: Sage.
Hanks, M., & Eckland, B. K. (1978). Adult voluntary associations and adolescent socialization. *Sociological Quarterly, 19*, 481–490.
Jennings, M. K. (2002). Generation units and the student protest movement in the United States: An intra- and intergenerational analysis. *Political Psychology, 23*, 303–324.
Jennings, M. K., Stoker, L., & Bowers, J. (1999, September). *Politics across generations: Family transmission reexamined*. Paper presented at the American Political Science Association convention, Atlanta, GA.
Mannheim, K. (1972). The problem of generations. In P. G. Altbach & R. S. Laufer (Eds.), *The new pilgrims* (pp. 101–135). New York: McKay. (Original work published 1928)
McAdam, D. (1988). *Freedom summer*. New York: Oxford University Press.
McAdam, D. (1992). Gender as a mediator of the activist experience: The case of freedom summer. *American Journal of Sociology, 97*, 1211–1240.
McLean, S. L., Schultz, D. A., & Steger, M. B. (Eds.). (2002). *Social capital: Critical perspectives on community and "bowling alone."* New York: New York University Press.
Putnam, R. (1995). Bowling alone: America's declining social capital. *Journal of Democracy, 6*, 65–78.
Putnam, R. (2000). *Bowling alone: The collapse and revival of American community*. New York: Simon & Schuster.
Rogler, L. (2002). Historical generations and psychology: The case of the Great Depression and World War II. *American Psychologist, 57*, 1013–1023.
Rosenblum, N. (1998). *Membership and morals*. Princeton, NJ: Princeton University Press.
Schuman, H., & Scott, J. (1989). Generations and collective memories. *American Sociological Review, 54*, 351–381.
Settersten, R. A. (1999). *Lives in time and place: The problems and promises of developmental science*. Amityville, NY: Baywood.
Sigel, R. (1996). *Ambition and accommodation: How women view gender relations*. Chicago: University of Chicago Press.
Snyder, R. C. (2002). Social capital: The politics of race and gender. In S. L. McLean, D. A. Schultz, & M. B. Steger (Eds.), *Social capital: Critical perspectives on community and "bowling alone"* (pp. 167–182). New York: New York University Press.
Stewart, A. J. (2003). Gender, race and generation in a midwest high school: Using ethnographically informed methods in psychology. *Psychology of Women Quarterly, 27*, 1–11.
Stewart, A. J., & Healy, J. M., Jr. (1986). The role of personality development and experience in shaping political commitment: An illustrative case. *Journal of Social Issues, 42*(2), 11–32.
Stewart, A. J., & Healy, J. M. (1989). Linking individual development and social changes. *American Psychologist, 44*, 30–42.
Stewart, A. J., Settles, I. H., & Winter, N. J. G. (1998). Women and the social movements of the 1960s: Activists, engaged observers, and nonparticipants. *Political Psychology, 19*, 63–94.
Sullivan, J. L., & Transue, J. E. (1999). The psychological underpinnings of democracy: A selective review of research on political tolerance, interpersonal trust, and social capital. *Annual Review of Psychology, 50*, 625–650.

Tangri, S. S., & Jenkins, S. R. (1993). The University of Michigan Class of 1967: The Women's Life Paths Study. In K. D. Hulbert & D. T. Schuster (Eds.), *Women's lives through time* (pp. 259–281). San Francisco: Jossey-Bass.

Winter, D. G., Stewart, A. J., Henderson-King, E., Henderson-King, D., & Lewis, A. E. (2003, July). *Predicting adult racial attitudes of Whites from high school interracial experience: A 40-year longitudinal test of the contact hypothesis.* Paper presented at the annual meeting of the International Society of Political Psychology, Boston.

Wuthnow, R. (1998). *Loose connections.* Cambridge, MA: Harvard University Press.

Yohalem, A. (1979). *The careers of professional women.* Montclair, NJ: Allanheld Osmun.

Yohalem, A. (1993). Columbia University graduate students, 1945–1951: The vanguard of professional women. In K. D. Hulbert & D. T. Schuster (Eds.), *Women's lives through time* (pp. 140–157). San Francisco: Jossey-Bass.

Youniss, J., McLellan, J. A., & Yates, M. (1997). What we know about engendering civic identity. *American Behavioral Scientist, 40,* 620–631.

Zucker, A. N. (1998). *Understanding feminist identity in three generations of college-educated women.* Unpublished doctoral dissertation, University of Michigan, Ann Arbor.

How Gene–Environment Interactions Shape Biobehavioral Development: Lessons From Studies With Rhesus Monkeys

Stephen J. Suomi
Laboratory of Comparative Ethology
National Institute of Child Health & Human Development
National Institutes of Health, DHHS

Recent research has found marked individual differences in patterns of rhesus monkey biobehavioral development throughout the life span. Approximately 20% of monkeys growing up in naturalistic settings consistently display unusually fearful and anxious-like behavioral reactions to novel, mildly stressful social situations throughout development; another 5%–10% are likely to exhibit impulsive and/or inappropriately aggressive responses under similar circumstances. These distinctive behavioral patterns and their biological correlates appear early in life and remain remarkably stable from infancy to adulthood. Both genetic and experiential mechanisms are implicated not only in the expression of these patterns but also in their transmission across successive generations of monkeys. For example, a specific polymorphism in the serotonin transporter gene is associated with deficits in infant neurobehavioral functioning and in juvenile and adolescent control of aggression and serotonin metabolism in monkeys that experienced insecure early attachments but not in monkeys that developed secure attachment relationships with their mothers during infancy (*maternal buffering*). Moreover, because the attachment style of a monkey mother is typically "copied" by her daughters when they grow up and become mothers themselves, similar buffering is likely to occur for the next generation of infants carrying that specific polymorphism.

The question of whether the features that make us unique as individuals are largely determined by our genetic heritage or shaped by our personal experi-

---

Requests for reprints should be sent to Stephen J. Suomi, Laboratory of Comparative Ethology, NICHD, NIH, DHHS, 6705 Rockledge Drive, Suite 8030, MSC 7971, Bethesda, MD 20892–7971. E-mail: suomis@lce.nichd.nih.gov

ences has been argued since at least the time of Aristotle; the nature–nurture debate clearly is not exactly new. What has been relatively new among researchers who study development is the realization that the basic question underlying this debate over the years may have been largely misdirected. Instead of arguing whether behavioral and biological characteristics that emerge during development are genetic in origin or are the product of specific experiences, these researchers now acknowledge that both genetic and environmental factors can play crucial roles in shaping individual developmental trajectories (Collins, Maccoby, Steinburg, Hetherington, & Bornstein, 2000). For example, behavioral geneticists (e.g., Plomin, 1990) have sought to determine the relative contributions of specific genetic and environmental factors to a variety of physical, physiological, behavioral, cognitive, and socioemotional features. Other investigators (e.g., Rutter, 2001) have focused on possible interactions between genetic and environmental factors.

It is unfortunate that direct study of potential gene–environment interactions and their possible influence on human development poses numerous methodological, practical, and even ethical problems. Demonstrations of specific gene–environment interactions are arguably most convincing when a range of predetermined genotypes can be studied prospectively across a range of systematically varied environments, yet that is almost never ethically proper and seldom practically feasible with human studies. Instead, such interactions must be largely inferred, typically after the fact. Specific genetic factors often tend to covary with particular rearing environments, such that the resulting gene–environment *correlations* can obscure actual gene–environment interactions (Reiss, Neiderhiser, Hetherington, & Plomin, 2000). Even when specific genes can be characterized in different individuals who are then followed prospectively in different environments, the actual basis for any emerging gene–environment interactions may be difficult to discern.

For example, Caspi et al. (2003) recently reported that the likelihood that individuals would experience episodes of depression was significantly related to the number of stressful life events they had experienced in the previous 5 years, but only if they possessed either the LS or SS variant of the serotonin transporter (5-HTT) gene; if they had the LL variant of the 5-HTT gene, there was no relation between recent depressive episodes and previous stressful life events. They also reported that individuals who had been maltreated as children were similarly more likely to experience depressive episodes in young adulthood than if they had not been maltreated, but again only if they possessed either the LS or SS (but not the LL) 5-HTT allele. However, Caspi et al. were unable to specify exactly what aspects of the stressful life events or childhood maltreatment might be responsible for these apparent gene–environment interactions. In addition, they acknowledged that

> This evidence that 5-HTTPR variation moderates the effect of life events on depression does not constitute unambiguous evidence of a G × E interaction, because exposure to life events may be influenced by genetic factors; if individuals have a heritable tendency to enter situation where they encounter stressful life events, these events may simply be a genetically saturated marker. (Caspi et al., 2003, p. 387)

Thus, even in this groundbreaking epidemiological study the lack of direct experimental control over environmental factors somewhat clouded precise interpretation of the extant findings.

Developmental researchers who study animals in captive environments are not faced with the same set of ethical and practical restrictions regarding manipulations of rearing environments characteristic of virtually all human studies. Instead, they can selectively breed or otherwise choose subjects with particular genetic pedigrees and rear them in different standardized environments in which various aspects of those environments can be systematically altered at various points throughout development. Such a research design readily permits the identification and characterization of specific gene–environment interactions (cf. Suomi, in press). Of course, the degree to which any findings from studies with animals can answer questions or address issues concerning human developmental phenomena is largely dependent on the degree to which the phenomena of interest generalize from the human case to the animal under study (Harlow, Suomi, & Gluck, 1972). In the case of many basic aspects of biological, behavioral, and socioemotional development, a growing body of evidence suggests considerable generality between humans and advanced nonhuman primate species.

In this article I summarize findings from a series of studies investigating genetic and environmental factors—and their interactions—that can shape individual differences in biobehavioral developmental trajectories in rhesus monkeys, especially with respect to the development of excessively fearful behavior, on the one hand, and overly aggressive behavior on the other. I begin with a description of the complex social contexts in which rhesus monkeys growing up in the wild normally encounter throughout development. Next, developmental trajectories for two subgroups of rhesus monkeys (*Macaca mulatta*)—one that consistently exhibit unusually fearful and anxious-like behavioral reactions to novel mildly stressful situations and another that are likely to exhibit impulsive and/or inappropriately aggressive responses in similar situations—are described in terms of their distinctive behavioral and biological features and propensities. Although many of these characteristics appear to be highly heritable, I present evidence demonstrating that they are also subject to considerable modification by environmental factors, especially those involving early social attachment relationships. Finally, I describe the results of studies investigating possible gene–environment interactions involving polymorphisms in the rhesus monkey 5-HTT gene that are structurally similar and functionally identical to the polymorphisms in the human 5-HTT gene cited earlier, and I discuss their relevance for considerations of human development.

SPECIES-NORMATIVE DEVELOPMENT IN RHESUS MONKEYS

Humans are not the only beings who grow up in a variety of complex social contexts. Most monkeys and apes living in natural settings spend their lives as active members of distinctive social communities, each typically characterized by elaborate kinship and status-defined social relationships. These primate communities often encompass three or more generations within individual family units and usually retain their basic identity long beyond the life span of any single community member or generation of members. Moreover, in most primate species the relationships between individual family members, between families within a given community, and even between different communities are dynamic in nature, and changes in each type of relationship often can have long-term consequences for all individuals involved.

An illustrative example of the complex and dynamic nature of social contexts within primate communities has been provided by extensive studies of rhesus monkeys in both captive and field settings over the last 50 years. Rhesus monkeys inhabit a wider geographic range, encompassing a broader mix of climate and habitat variation, than any other nonhuman primate species, with perhaps one or two exceptions. In contrast to many primate species currently classified as endangered or threatened, rhesus monkeys are actually expanding their local populations in certain parts of their extensive range. They have also consistently demonstrated an impressive ability to adapt to and indeed thrive in a wide variety of captive environments (Novak & Suomi, 1991).

In nature, rhesus monkeys reside in large social groups (troops), each comprised of several different female-headed families (matrilines) spanning several generations of kin, plus numerous immigrant males. This pattern of social organization derives from the fact that rhesus monkey females stay in their natal troop for their entire lives, whereas virtually all rhesus monkey males emigrate from their natal troop around the time of puberty, usually in their 4th or 5th year, and then join other troops (Lindburg, 1971). These troops are also characterized by multiple social dominance relationships, including distinctive hierarchies both between and within families as well as a hierarchy among the immigrant adult males, whose relative status seems to be largely a function of their ability to join and maintain coalitions, especially with high-ranking females. Indeed, the dominance status of any particular rhesus monkey within its troop depends not so much on how big and strong it is but rather who its family and friends are—and the latter is clearly dependent on the development of complex social skills during ontogeny.

Given the multiple social relationships and the variety of social networks embedded within each rhesus monkey troop, how do infants born to troop members become integrated into the overall social fabric of that troop? An impressive body of both laboratory and field data suggests that such integration is an emergent consequence of the species-normative pattern of socialization that rhesus monkey infants experience as they grow up (Sameroff & Suomi, 1996). In particular, the social networks of rhesus monkey infants initially are largely limited to members of their immediate families but expand dramatically in both scope and complexity as these infants mature.

Rhesus monkey infants begin life completely dependent on their mother for all their essential needs. They spend virtually all of their first month of life in physical contact with or within arm's reach of their mother, and mothers typically limit any other social contact of their infants to female members of their immediate family (Harlow, Harlow, & Hansen, 1963; Hinde & Spencer-Booth, 1967). During this time, a strong and enduring social bond inevitably develops between mother and infant, recognized by Bowlby (1969) to be basically homologous with the mother–infant attachment relationship universally seen in all human cultures.

Once infants have established an attachment bond with their mother, they quickly learn to use her as a secure base from which to start exploring their environment, beginning as early as their second month of life. Shortly thereafter they spend increasing amounts of time engaging in social interactions with other troop members, especially peers. Although mothers show considerable variability in the degree to which they permit, if not encourage, these interactions with peers, by 6 months of age most youngsters typically are spending many hours each week in peer-directed activities. Interactions with peers continue to increase in both frequency and complexity throughout the rest of the young monkeys' first year of life. In contrast, the amount of time they spend interacting with their mother declines substantially after weaning, and this decline typically accelerates if the mother becomes pregnant again (Berman, Rasmussen, & Suomi, 1993).

Peer-directed activities continue at high and essentially stable rates throughout the 2nd and 3rd years of life (Ruppenthal, Harlow, Eisele, Harlow, & Suomi, 1974). During this time, peer play becomes increasingly gender specific and sex segregated (i.e., males tend to play more with males and females with females) and involves behavioral sequences that appear to simulate virtually all adult social activities, including courtship and reproductive behaviors as well as dominance–aggressive interactions. However, both male and female juveniles still retain unique social ties with their mother even as she becomes increasingly involved with their younger siblings. They continue to use her as a secure base, they routinely seek physical contact with her under stressful circumstances, and they actively solicit her participation (and respond to her solicitations) in agonistic exchanges with other monkeys both inside and outside their troop (Suomi, 1998).

The social activities of and contexts for male and female juveniles change dramatically and differentially with puberty, which usually begins at the end of the 3rd year for females and the start of the 4th year for males. Although females remain in their natal troop throughout adolescence and thereafter, their interactions with peers decline substantially from prepubertal levels as they redirect many of their social activities toward matrilineal kin, including both their mothers and the offspring they subsequently bear and rear. Nevertheless, females do retain some aspects of previous relationships with female peers throughout adult life, although their mutual activities tend to be greatly diminished in frequency and are largely limited to grooming bouts and (paradoxically) agonistic encounters rather than play. Much more time is instead invested in family-directed activities; indeed, females remain actively involved in family affairs for the rest of their lives, even after they stop having babies themselves.

Pubertal males, by contrast, leave both their family and their natal troop permanently, typically joining all-male "gangs" for varying periods before attempting to enter a different troop. In doing so, these young males effectively terminate their relationship with their mother and all other female relatives, inasmuch as they are not permitted to re-enter their natal troop once they have emigrated. They also end their relationships with most of their natal group peers (Suomi, 1998). Once a male has joined a new troop he must not only establish new relationships with the various members but also learn about the specific kinship relationships and multiple dominance hierarchies in order to become successfully integrated within that troop. Not surprisingly, this period of transition represents a time of major stress and potential danger for adolescent and young adult males: The mortality rate for males during the process of natal troop emigration and subsequent immigration approaches 50% in some wild monkey populations (Dittus, 1979). Some surviving males remain in their new troop for the rest of their lives, whereas other males may transfer from one troop to another several times during their adult years, but they never return to their natal troop (Berard, 1989).

INDIVIDUAL DIFFERENCES IN THE REGULATION OF FEAR AND AGGRESSION

Although virtually all rhesus monkeys growing up both in the wild and in captive social groups maintained in captivity go through the same basic developmental sequences just described, the specific social experiences that any individual accrues, and the specific contexts in which they occur throughout development, can vary considerably both within and between families. For example, numerous studies have demonstrated that most rhesus monkey mothers are acutely sensitive to those aspects of their immediate physical and social environment that pose potential threats to their infant's well-being, and they appear to adjust their maternal behavior accordingly. Both laboratory and field studies have consistently shown that low-ranking mothers tend to be much more restrictive of their infant's exploratory efforts than are high-ranking mothers, whose own maternal style has been termed "laissez-faire" (Altmann, 1980). The standard interpretation of these findings is that low-ranking mothers risk reprisal from others if they try to intervene whenever their infant is threatened, so they minimize such risk by restricting their infant's exploration. High-ranking mothers have no such problem; hence, they can afford to let their infant explore as it pleases.

Other studies have found that mothers generally become more restrictive and increase their levels of infant monitoring when their immediate social environment becomes less stable, such as when major changes in interfamily dominance hierarchies take place or when alien males try to join the troop. Changes in various aspects of the physical environment, such as the food supply becoming less predictable, have also been associated with increases in maternal restriction of early infant environmental exploration (Andrews & Rosenblum, 1991), as have increases in the population of any one troop (Berman, Rasmussen, & Suomi, 1997). For infants whose opportunities to explore are chronically limited during their first few months of life, development of species-normative relationships with others in their social group, especially peers, can be compromised, often with long-term consequences for both the infants and their social group (Suomi, 1997).

There are also substantial differences among individual troop members in the precise timing and relative ease with which they make major developmental transitions as well as with how they manage the day-to-day challenges and stresses that are an inevitable consequence of complex social group life. In particular, recent research has identified two subgroups of monkeys that exhibit specific problems in their socioemotional regulation that can result in increased long-term risk for behavioral pathology and even mortality. Members of one subgroup, comprising approximately 15% to 20% of both wild and captive populations, seem excessively fearful. These monkeys consistently respond to novel and/or mildly challenging situations with extreme behavioral disruption and pronounced physiological arousal (Suomi, 1986).

Highly fearful monkeys can usually be identified during their initial weeks and months of life. Most begin leaving their mothers later chronologically and explore their physical and social environment less than other infants in their birth cohort. Highly fearful youngsters also tend to be shy and withdrawn in their initial encounters with peers; laboratory studies have shown that they exhibit significantly higher and more stable heart rates and greater secretion of cortisol in such interactions than do their less reactive age-mates. However, when these fearful monkeys are in familiar and stable social settings they are virtually indistinguishable, both behaviorally and physiologically, from their peers (Suomi, 1991b). In contrast, when fearful monkeys encounter extreme and/or prolonged stress, their behavioral and physiological differences from others in their social group become even more pronounced.

For example, young rhesus monkeys typically experience functional maternal separations during the 2-month-long annual breeding season when their mothers repeatedly leave the troop for brief periods to consort with selected males (Berman, Rasmussen, & Suomi, 1994). The sudden loss of access to its mother is a major social stressor for any young monkey and, not surprisingly, virtually all youngsters initially react to their mother's departure with short-term behavioral agitation and physiological arousal, much as Bowlby (1973) described for human infants experiencing involuntary maternal separation. However, whereas most youngsters soon begin to adapt to the separation and readily seek out the company of others in their social group until their mother returns, highly fearful individuals typically lapse into a behavioral depression characterized by increasing lethargy, lack of apparent interest in social stimuli, eating and sleeping difficulties, and a characteristic hunched-over, fetal-like posture (Suomi, 1991a).

Laboratory studies simulating these naturalistic maternal separations have shown that, relative to their like-reared peers, highly fearful monkeys not only are more likely to exhibit depressive-like behavioral reactions to short-term social separation but also tend to show greater and more prolonged hypothalamic–pituitary–adrenal (HPA) activation, more dramatic sympathetic arousal, more rapid central noradrenergic turnover, and greater immunosuppression (Suomi, 1991b). These differential patterns of biobehavioral response to separation tend to remain remarkably stable throughout prepubertal development and may be maintained during adolescence and even into adulthood (Suomi, 1995). Moreover, individual differences in infant biobehavioral response to separation are predictive of differential responses to other situations experienced later in life (e.g., Fahlke et al., 2000). An increasing body of evidence has demonstrated significant heritability for at least some components of these differential responses to stress (Higley et al., 1993; Williamson et al., 2003).

Highly fearful males usually emigrate from their natal troop at significantly older ages than the rest of their male birth cohort and, when they do finally leave, they typically use much more conservative strategies for entering a new troop than do their less fearful peers. This pattern of delayed emigration may actually be adaptive, in that the larger physically and heavier a male is at the time he emigrates from his natal troop, the greater the likelihood that he will survive and successfully join another troop (Rasmussen, Fellows, & Suomi, 1990). Therefore, if a male is able to postpone emigration until he has largely finished his adolescent growth spurt, he appears to be better able to make the transition to adult male life than if he leaves home prior to or during the growth spurt. Because fearful adolescent males pose little apparent threat to adult females and their offspring, they tend to be tolerated by other troop members at ages when the other males in their birth cohort have either left voluntarily or been forcibly driven away. Thus, even though excessive fearfulness apparently puts an individual male at increased risk for adverse biobehavioral reactions to stress throughout development, there may be some circumstances in which this characteristic can actually be adaptive (Suomi, 2000b).

A parallel situation exists for females: Highly fearful young mothers in the wild tend to reject and punish their infants at higher rates around the time of weaning than do other mothers in their troop (Rasmussen, Timme, & Suomi, 1997), and in the absence of social support they appear to be at increased risk for infant neglect and/or abuse (Suomi & Ripp, 1983). Yet under stable social circumstances these fearful females may not only turn out to be highly competent mothers but also often achieve relatively high positions of social dominance (Rasmussen et al., 1997). In sum, excessive fearfulness in infancy appears to be associated with increased risk for developing anxious- and depressive-like symptoms and potential problems in parenting in response to stressful circumstances later in life, but such long-term outcomes are far from inevitable.

Other rhesus monkeys appear to have problems regulating aggression. These monkeys, comprising approximately 5% to 10% of the population, seem unusually impulsive, insensitive, and overly aggressive in their interactions with other troop members. Impulsive young males seem unable to moderate their responses to rough-and-tumble play initiations from peers, frequently escalating initially benign play bouts into full-blown aggressive exchanges that may result in actual wounding (Higley, Mehlman, Taub, et al., 1992). Not surprisingly, most of these individuals tend to be avoided by peers, and as a result they become increasingly isolated socially. In addition, many of these juvenile males often appear unwilling (or unable) to follow the "rules" inherent in rhesus monkey social dominance hierarchies. For example, they may directly challenge a dominant adult male, a foolhardy act that can result in serious injury, especially when the juvenile refuses to back away or exhibit submissive behavior once defeat becomes obvious. Impulsive males also show a propensity for making dangerous leaps from treetop to treetop, sometimes with injurious or even fatal outcomes (Mehlman et al., 1994).

Overly impulsive monkeys of both genders consistently exhibit chronic deficits in central serotonin metabolism, as reflected by unusually low cerebrospinal fluid (CSF) concentrations of the primary central serotonin metabolite 5-hydroxyindoleacetic acid (5-HIAA). Laboratory studies have shown that these deficits in serotonin metabolism appear early in life and tend to persist throughout development, as was the case for HPA responsiveness among highly fearful monkeys. Monkeys who exhibit such deficits are also likely to show poor state control and visual orienting capabilities during early infancy (Champoux, Suomi, & Schneider, 1994), poor performance on delay-of-gratification tasks as juveniles (Bennett et al., 1999), and sleep disturbances as adults (Zajicek, Higley, Suomi, &

Linnoila, 1997). Moreover, individual differences in 5-HIAA concentrations appear to be highly heritable among monkeys of similar age and comparable rearing background (Higley et al., 1993).

Recent field studies have found that the timing of natal troop emigration typical for impulsive males is seemingly the reverse of that for fearful males, often with deadly consequences. Ostracized by their peers and frequently attacked by adults of both sexes, most of these excessively aggressive young males are physically driven out of their natal troop prior to 3 years of age, well before the onset of puberty and long before most of their male cohort begins the normal emigration process (Mehlman et al., 1995). These males tend to be grossly incompetent socially and, lacking the requisite social skills necessary for successful entrance into another troop or even into an all-male gang, most of them become solitary and typically perish within a year (Higley, Mehlman, et al., 1996).

Young females who have chronically low CSF levels of 5-HIAA also tend to be impulsive, aggressive, and generally rather incompetent socially. However, unlike the males, they are not expelled from their natal troop but instead remain with their family throughout their lifetime, although studies of captive rhesus monkey groups suggest that these females will likely remain at the bottom of their respective dominance hierarchies (Higley, King, et al., 1996). Although most become mothers, their maternal behavior is often inadequate or even abusive, such that the attachment relationships they develop with their offspring tend to be avoidant, if not disorganized (Suomi, 2000a). In sum, rhesus monkeys who exhibit poor regulation of impulsive and aggressive behavior and low central serotonin metabolism early in life tend to follow developmental trajectories that often result in premature death for males and chronically low social dominance and poor parenting for females.

EFFECTS OF PEER-REARING ON RHESUS MONKEY BIOBEHAVIORAL DEVELOPMENT

Although the findings from both the field and laboratory studies cited thus far have shown that individual differences in expressions of fear and aggression tend to be quite stable from infancy to adulthood and are at least in part heritable, this does not mean that they are necessarily fixed at birth, immune to subsequent environmental influence. To the contrary, an increasing body of evidence from laboratory studies has demonstrated that patterns of socioemotional development, neuroendocrine responsiveness, and neurotransmitter metabolism alike can be modified substantially by certain early social experiences, especially those involving early attachment relationships.

Some of the most compelling evidence comes from studies of rhesus monkey infants raised with peers instead of their biological mothers. During their initial weeks of life, peer-reared infants readily establish strong social bonds with each other, much as mother-reared infants develop attachments to their own mothers (Harlow, 1969). However, because peers are not nearly as effective as typical monkey mothers in reducing fear in the face of novelty or in providing secure bases for environmental exploration, the attachment relationships that these peer-reared infants develop are almost always "anxious" in nature (Suomi, 1995). As a result, although peer-reared monkeys show completely normal physical and motor development, most appear to be excessively fearful: Their early exploratory behavior tends to be somewhat limited, they seem reluctant to approach novel objects, and they tend to be shy in initial encounters with unfamiliar peers (Suomi, 1991b).

Social play among peer-reared monkeys occurs less frequently and tends to be less complex with respect to the duration, the diversity of specific behaviors incorporated, and the number of partners involved in most play episodes than is typically shown by their mother-reared counterparts. One explanation for their relatively poor play performance is that their partners have to serve both as attachment figures and playmates, dual roles that neither mothers nor mother-reared peers have to fulfill. Another obstacle to developing sophisticated play repertoires faced by peer-reared monkeys is that all of their early play bouts involve partners who are basically as incompetent socially as they are. Perhaps as a result of these factors, peer-reared youngsters typically drop to the bottom of their respective dominance hierarchies when they are subsequently housed with mother-reared monkeys their own age (Higley, Suomi, & Linnoila, 1996).

Several prospective longitudinal studies have found that peer-reared monkeys consistently exhibit more extreme behavioral, adrenocortical, and noradrenergic reactions to social separations than do their mother-reared cohorts, even after they have been living in the same social groups for extended periods (e.g., Higley & Suomi, 1989; Higley, Suomi, & Linnoila, 1992). Such differences in prototypical biobehavioral reactions to separation persist from infancy to adolescence, if not beyond. It is interesting that the general nature of the separation reactions exhibited by peer-reared monkeys seems to mirror that shown by "naturally occurring" highly fearful mother-reared subjects. In this sense, early rearing with peers appears to have the effect of making rhesus monkey infants generally more fearful than they might have been if reared by their biological mothers (Suomi, 1991b).

Early peer rearing has another long-term developmental consequence for rhesus monkeys: It tends to make them more impulsive, especially if they are males. Like the previously described impulsive monkeys growing up in the wild, peer-reared males initially exhibit overly aggressive tendencies in the context of juvenile play; as they approach puberty, the frequency and severity of their aggressive episodes typically exceed those of mother-reared group members of similar age. Peer-reared females tend to groom (and be groomed by) others in their social group less frequently and for shorter durations than their mother-reared counterparts, and they usually stay at the bottom of their respective dominance hierarchies. These differences between peer-reared and mother-reared age mates in aggression, grooming, and dominance remain relatively robust throughout the preadolescent and adolescent years (Higley, Suomi, & Linnoila, 1996). Peer-reared monkeys also consistently show lower CSF concentrations of 5-HIAA than their mother-reared counterparts. These group differences in 5-HIAA concentrations appear well before 6 months of age, and they remain stable at least throughout adolescence and into early adulthood (Higley & Suomi, 1996). Thus, peer-reared monkeys exhibit the same general tendencies that characterize excessively impulsive wild-living (and mother-reared) rhesus monkeys, not only behaviorally but also in terms of decreased serotonergic functioning.

GENE–ENVIRONMENT INTERACTIONS

Studies examining the effects of peer rearing and other variations in early rearing history (e.g., Harlow & Harlow, 1969), along with the previously cited heritability findings, clearly provide compelling evidence that both genetic and early experiential factors can affect a monkey's capacity to regulate expression of fear and aggression. Do these factors operate independently, or do they interact in some fashion in shaping individual developmental trajectories? Ongoing research capitalizing on the discovery of polymorphisms in one specific gene—the serotonin transporter gene—suggests that gene–environment interactions not only occur but also can be expressed in multiple forms.

The serotonin transporter gene (5-HTT), considered by biological psychiatrists to be a "candidate" gene for impaired serotonergic function (Lesch et al., 1996), has length variation in its promoter region that results in allelic variation in serotonin expression. A heterozygous "short" allele (LS) confers low transcriptional efficiency to the 5-HTT promoter relative to the homozygous "long" allele (LL), raising the possibility that low 5-HTT expression may result in decreased serotonergic functioning (Heils et al., 1996), although evidence in support of this hypothesis in humans has been decidedly mixed to date (e.g., Furlong et al., 1998). The 5-HTT polymorphism was first characterized in humans but, as previously mentioned, it also appears in a largely homologous form in rhesus monkeys—but, interestingly, *not* in many other species of primates and not in any other mammals studied to date (Lesch et al., 1997).

My colleagues and I (Suomi, 2004) recently used polymerase chain reaction techniques to characterize the 5-HTT allelic status of monkeys used in the studies comparing peer-reared monkeys with mother-reared controls described earlier. Because extensive observational data and biological samples had been previously collected from these monkeys throughout development, it was possible to examine a wide range of behavioral and physiological measures for potential 5-HTT polymorphism interactions with early rearing history. Analyses completed to date suggest that such interactions are widespread and diverse.

For example, Bennett et al. (2002) found that CSF 5-HIAA concentrations did not differ as a function of 5-HTT status for mother-reared subjects, whereas among peer-reared monkeys individuals with the LS allele had significantly lower CSF 5-HIAA concentrations than those with the LL allele. One interpretation of this interaction is that mother-rearing appeared to "buffer" any potentially deleterious effects of the LS allele on serotonin metabolism. Barr, Newman, Becker, et al. (in press) reported a similar buffering effect with respect to aggression: High levels of aggression were found in peer-reared monkeys with the LS allele, whereas mother-reared LS monkeys exhibited low levels that were comparable to those of both mother- and peer-reared LL monkeys. In addition, Barr, Newman, Shannon, et al. (in press) found a parallel pattern of maternal buffering with respect to HPA activity in response to social separation: LS peer-reared juveniles had significantly higher ACTH concentrations than LS mother-reared, LL mother-reared, and LL peer-reared juveniles. Finally, Champoux et al. (2002) examined the relation between early rearing history and serotonin transporter gene polymorphic status on measures of neonatal neurobehavioral development during the first month of life and found further evidence of maternal buffering. Specifically, infants possessing the LS allele that were being reared in the laboratory neonatal nursery showed significant deficits in measures of attention, activity, and motor maturity relative to nursery-reared infants possessing the LL allele, whereas both LS and LL infants being reared by competent mothers exhibited normal values for each of these measures.

In sum, the consequences of having the LS allele differed dramatically for peer-reared and mother-reared monkeys. Peer-reared individuals with the LS allele exhibited deficits in measures of neurobehavioral functioning during their initial weeks of life, increased HPA activity and high levels of aggression as juveniles, and reduced serotonin metabolism as adolescents. Mother-reared subjects with the very same allele developed normal early neurobehavioral functioning, HPA activity, regulation of aggression, and serotonin metabolism. Indeed, one could argue on the basis of these findings that having the LS allele may well lead to psychopathology among monkeys with poor early rearing histories but might actually be adaptive for monkeys who develop secure early attachment relationship with their mothers.

The implications of these recent findings are considerable with respect to the potential for cross-generational transmission of these biobehavioral characteristics. Among the most intriguing aspects of the long-term consequences of different early attachment experiences is the apparent transfer of specific features of maternal behavior across successive generations. Several studies of rhesus monkeys and other nonhuman primate species have demonstrated strong continuities between the type of attachment relationship a female infant develops with her mother and the type of attachment relationship she develops with her own infant(s) when she becomes a mother herself (Suomi, 1999). In particular, the pattern of ventral contact a female infant has with her mother (or mother substitute) during her initial months of life is a powerful predictor of the pattern of ventral contact she will have with her own infants during their first 6 months of life (Champoux, Byrne, Delizio, & Suomi, 1992; Fairbanks, 1989). This predictive cross-generational relationship is as strong in females who were foster reared from birth by unrelated multiparous females as it is for females reared by their biological mothers, strongly suggesting that such cross-generational transmission necessarily involves nongenetic mechanisms (Suomi & Levine, 1998).

If similar maternal buffering is indeed experienced by the next generation of infants carrying the LS 5-HTT polymorphism, then having had their mothers develop a secure attachment relationship with their own mothers when they were infants themselves might well provide the basis for a nongenetic means of transmitting its apparently adaptive consequences to that new generation of monkeys. On other hand, if contextual factors, such as changes in maternal dominance rank, instability within the troop, or changes in the availability of food, were to alter a young mother's care of her infants in a way that compromised any such buffering, then one might well expect any offspring carrying the LS 5-HTT polymorphism to develop some of the problems described earlier. What the relevant nongenetic mechanisms might be, and through what developmental processes might they act, are questions at the heart of ongoing investigations.

My colleagues and I are currently carrying out parallel studies focusing on other potential gene–environment interactions involving not only 5-HTT polymorphisms but also polymorphisms in other "candidate" genes, such as MAO-A, and our findings to date suggest that such interactions are ubiquitous and can occur at a variety of points throughout development. We are also currently involved in collaborative studies investigating patterns of gene expression in different brain regions and how such patterns might be influenced by early experience. At the very least, our findings suggest that the social context in which a rhesus monkey infant is reared can have far-ranging consequences throughout the whole of development—not only at the levels of behavioral functioning and socioemotional regulation but also at the levels of neurohormonal responsiveness, neurotransmitter metabolism, and perhaps even gene expression. The context in which development takes place clearly matters a great deal for rhesus monkeys. It is hard to imagine how it could be less so for human development.

REFERENCES

Altmann, J. (1980). *Baboon mothers and infants*. Cambridge, MA: Harvard University Press.

Andrews, M. W., & Rosenblum, L. A. (1991). Security of attachment in infants raised in variable or low-demand environments. *Child Development, 62,* 686–693.

Barr, C. S., Newman, T. K., Becker, M. L., Parker, C. C., Champoux, M., Lesch, K. P., et al. (in press). Early experience and rh5-HTTPLR genotype interact to influence social behavior and aggression in nonhuman primates. *Genes, Brain, and Behavior.*

Barr, C. S., Newman, T. K., Shannon, C., Parker, C., Dvoskin, R. L., Becker, M. L., et al. (in press). Rearing condition and rh5-HTTLPR interact to influence LHPA axis response to stress in infant monkeys. *Biological Psychiatry.*

Bennett, A. J., Lesch, K. P., Heils, A., Long, J., Lorenz, J., Shoaf, S. E., et al. (2002). Early experience and serotonin transporter gene variation interact to influence primate CNS function. *Molecular Psychiatry, 17,* 118–122.

Bennett, A. J., Tsai, T., Hopkins, W. D., Lindell, S. G., Pierre, P. J., Champoux, M., & Shoaf, S. E. (1999). Early social environment influences acquisition of a computerized joystick task in rhesus monkeys (*Macaca mulatta*). *American Journal of Primatology, 49,* 33–34.

Berard, J. (1989). Male life histories. *Puerto Rican Health Sciences Journal, 8,* 47–58.

Berman, C. M., Rasmussen, K. L. R., & Suomi, S. J. (1993). Reproductive consequences of maternal care patterns during estrus among free-ranging rhesus monkeys. *Behavioral Ecology and Sociobiology, 32,* 391–399.

Berman, C. M., Rasmussen, K. L. R., & Suomi, S. J. (1994). Responses of free-ranging rhesus monkeys to a natural form of maternal separation: I. Parallels with mother–infant separation in captivity. *Child Development, 65,* 1028–1041.

Berman, C. M., Rasmussen, K. L. R., & Suomi, S. J. (1997). Group size, infant development, and social networks: A natural experiment with free-ranging rhesus monkeys. *Animal Behavior, 53,* 405–421.

Bowlby, J. (1969). *Attachment.* New York: Basic Books.

Bowlby, J. (1973). *Separation.* New York: Basic Books.

Caspi, A., Sugen, K., Moffitt, T. E., Taylor, A., Craig, I. W., Harrington, H., et al. (2003). Influence of life stress on depression: Moderation by a polymorphism in the 5-HTT gene. *Science, 301,* 386–389.

Champoux, M., Bennett, A. J., Lesch, K. P., Heils, A., Nielson, D. A., Higley, J. D., & Suomi, S. J. (2002). Serotonin transporter gene polymorphism and neurobehavioral development in rhesus monkey neonates. *Molecular Psychiatry, 7,* 1058–1063.

Champoux, M., Byrne, E., Delizio, R. D., & Suomi, S. J. (1992). Motherless mothers revisited: Rhesus maternal behavior and rearing history. *Primates, 33,* 251–255.

Champoux, M., Suomi, S. J., & Schneider, M. L. (1994). Temperamental differences between captive Indian and Chinese–Indian hybrid rhesus macaque infants. *Laboratory Animal Science, 44,* 351–357.

Collins, W. A., Maccoby, E. E., Steinburg, L., Hetherington, E. M., & Bornstein, M. H. (2000). Contemporary research on parenting: The case for nature *and* nurture. *American Psychologist, 55,* 218–232.

Dittus, W. P. J. (1979). The evolution of behaviours regulating density and age-specific sex ratios in a primate population. *Behaviour, 69,* 265–302.

Fahlke, C., Lorenz, J. G., Long, J., Champoux, M., Suomi, S. J., & Higley, J. D. (2000). Rearing experiences and stress-induced plasma cortisol as early risk factors for excessive alcohol consumption in nonhuman primates. *Alcoholism: Clinical and Experimental Research, 24,* 644–650.

Fairbanks, L. A. (1989). Early experience and cross-generational continuity of mother–infant contact in vervet monkeys. *Developmental Psychobiology, 22,* 669–681.

Furlong, R. A., Ho, L., Walsh, C., Rubinsztein, J. S., Jain, S., Pazkil, E. S., et al. (1998). Analysis and meta-analysis of two serotonin transporter gene polymorphisms in bipolar and unipolar affective disorders. *American Journal of Medical Genetics, 81,* 58–63.

Harlow, H. F. (1969). Age-mate or peer affectional system. In D. S. Lehrman, R. A. Hinde, & E. Shaw (Eds.), *Advances in the study of behavior* (Vol. 2, pp. 333–383). New York: Academic.

Harlow, H. F., & Harlow, M. K. (1969). Effects of various mother–infant relationships on rhesus monkey behaviors. In B. M. Foss (Ed.), *Determinants of infant behaviour* (Vol. 4, pp. 15–36). London: Methuen.

Harlow, H. F., Harlow, M. K., & Hansen, E. W. (1963). The maternal affectional system of rhesus monkeys. In H. L. Rheingold (Ed.), *Maternal behavior in mammals* (pp. 254–281). New York: Wiley.

Harlow, H. F., Suomi, S. J., & Gluck, J. P. (1972). Generalization of behavioral data between nonhuman and human animals. *American Psychologist, 27,* 709–716.

Heils, A., Teufel, A., Petri, S., Stober, G., Riederer, P., Bengel, B., & Lesch, K. P. (1996). Allelic variation of human serotonin transporter gene expression. *Journal of Neurochemistry, 6,* 2621–2624.

Higley, J. D., King, S. T., Hasert, M. F., Champoux, M., Suomi, S. J., & Linnoila, M. (1996). Stability of individual differences in serotonin function and its relationship to severe aggression and competent social behavior in rhesus macaque females. *Neuropsychopharmacology, 14,* 67–76.

Higley, J. D., Mehlman, P. T., Taub, D. M., Higley, S., Fernald, B., Vickers, J. H., et al. (1996). Excessive mortality in young free-ranging male nonhuman primates with low CSF 5-HIAA concentrations. *Archives of General Psychiatry, 53,* 537–543.

Higley, J. D., Mehlman, P. T., Taub, D. M., Higley, S. B., Vickers, J. H., Suomi, S. J., & Linnoila, M. (1992). Cerebrospinal fluid moamine and adrenal correlates of aggression in free-ranging rhesus monkeys. *Archives of General Psychiatry, 49,* 436–444.

Higley, J. D., & Suomi, S. J. (1989). Temperamental reactivity in nonhuman primates. In G. A. Kohnstamm, J. E. Bates, & M. K. Rothbard (Eds.), *Handbook of temperament in children* (pp. 153–167). New York: Wiley.

Higley, J. D., & Suomi, S. J. (1996). Reactivity and social competence affect individual differences in reaction to severe stress in children: Investigations using nonhuman primates. In C. R. Pfeffer (Ed.), *Intense stress and mental disturbance in children* (pp. 3–58). Washington, DC: American Psychiatric Press.

Higley, J. D., Suomi, S. J., & Linnoila, M. (1992). A longitudinal assessment of CSF monoamine metabolite and plasma cortisol concentrations in young rhesus monkeys. *Biological Psychiatry, 32,* 127–145.

Higley, J. D., Suomi, S. J., & Linnoila, M. (1996). A nonhuman primate model of Type II alcoholism?: Part 2: Diminished social competence and excessive aggression correlates with low CSF 5-HIAA concentrations. *Alcoholism: Clinical and Experimental Research, 20,* 643–650.

Higley, J. D., Thompson, W. T., Champoux, M., Goldman, D., Hasert, M. F., Kraemer, G. W., et al. (1993). Paternal and maternal genetic and environmental contributions to CSF monoamine metabolites in rhesus monkeys (*Macaca mulatta*). *Archives of General Psychiatry, 50,* 615–623.

Hinde, R. A., & Spencer-Booth, Y. (1967). The behaviour of socially living rhesus monkeys in their first two and a half years. *Animal Behaviour, 15,* 169–176.

Lesch, K. P., Bengel, D., Heils, A., Sabol, S. Z., Greenberg, B. D., Petri, S., et al. (1996). Association of anxiety-related traits with a polymorphism in the serotonin transporter gene regulatory region. *Science, 274,* 1527–1531.

Lesch, L. P., Meyer, J., Glatz, K., Flugge, G., Hinney, A., Hebebrand, J., et al. (1997). The 5-HT transporter gene-linked polymorphic region (5-HTTLPR) in evolutionary perspective: Alternative biallelic variation in rhesus monkeys. *Journal of Neural Transmission, 104,* 1259–1266.

Lindburg, D. G. (1971). The rhesus monkey in north India: An ecological and behavioral study. In L. A. Rosenblum (Ed.), *Primate behavior: Developments in field and laboratory research* (Vol. 2, pp. 1–106). New York: Academic.

Mehlman, P. T., Higley, J. D., Faucher, I., Lilly, A. A., Taub, D. M., Vickers, J. H., et al. (1994). Low cerebrospinal fluid 5 hydroxyindoleacetic acid concentrations are correlated with severe aggression and reduced impulse control in free-ranging primates. *American Journal of Psychiatry, 151,* 1485–1491.

Mehlman, P. T., Higley, J. D., Faucher, I., Lilly, A. A., Taub, D. M., Vickers, J. H., et al. (1995). CSF 5-HIAA concentrations are correlated with sociality and the timing of emigration in free-ranging primates. *American Journal of Psychiatry, 152,* 901–913.

Novak, M. A., & Suomi, S. J. (1991). Social interaction in nonhuman primates: An underlying theme for primate research? *Laboratory Animal Science, 41,* 308–314.

Plomin, R. (1990). *Nature and nurture: An introduction to human behavioral genetics.* Pacific Grove, CA: Brooks/Cole.

Rasmussen, K. L. R., Fellows, J. R., & Suomi, S. J. (1990). Physiological correlates of emigration behavior and mortality in adolescent male rhesus monkeys on Cayo Santiago. *American Journal of Primatology, 20,* 224–225.

Rasmussen, K. L. R., Timme, A., & Suomi, S. J. (1997). Comparison of physiological measures of Cayo Santiago rhesus monkey females within and between social groups. *Primate Reports, 47,* 49–55.

Reiss, D., Neiderhiser, J. M., Hetherington, E. M., & Plomin, R. (2000). *The relationship code: Deciphering genetic and social influences on adolescent development.* Cambridge, MA: Harvard University Press.

Ruppenthal, G. C., Harlow, M. K., Eisele, C. D., Harlow, H. F., & Suomi, S. J. (1974). Development of peer interactions of monkeys reared in a nuclear family environment. *Child Development, 45,* 670–682.

Rutter, M. (2001). How can we know environment really matters? In F. Lamb-Parker, J. Hagen, & R. Robinson (Eds.), *Developmental and contextual transition of children and families: Implications for research, policy, & practice* (pp. 3–18). New York: Columbia University Press.

Sameroff, A. J., & Suomi, S. J. (1996). Primates and persons: A comparative developmental understanding of social organization. In R. B. Cairns, G. H. Elder, & E. J. Costello (Eds.), *Developmental science* (pp. 97–120). Cambridge, England: Cambridge University Press.

Suomi, S. J. (1986). Anxiety-like disorders in young primates. In R. Gittelman (Ed.), *Anxiety disorders of childhood* (pp. 1–23). New York: Guilford.

Suomi, S. J. (1991a). Primate separation models of affective disorders. In J. Madden (Ed.), *Neurobiology of learning, emotion, and affect* (pp. 195–214). New York: Raven.

Suomi, S. J. (1991b). Up-tight and laid-back monkeys: Individual differences in the response to social challenges. In S. Brauth, W. Hall, & R. Dooling (Eds.), *Plasticity of development* (pp. 27–56). Cambridge, MA: MIT Press.

Suomi, S. J. (1995). Influence of Bowlby's attachment theory on research on nonhuman primate biobehavioral development. In S. Goldberg, R. Muir, & J. Kerr (Eds.), *Attachment theory: Social, developmental, and clinical perspectives* (pp. 185–201). Hillsdale, NJ: Analytic Press.

Suomi, S. J. (1997). Early determinants of behaviour: Evidence from primate studies. *British Medical Bulletin, 53,* 170–184.

Suomi, S. J. (1998). Conflict and cohesion in rhesus monkey family life. In M. Cox & J. Brooks-Gunn (Eds.), *Conflict and cohesion in families* (pp. 283–296). Mahwah, NJ: Lawrence Erlbaum Associates, Inc.

Suomi, S. J. (1999). Attachment in rhesus monkeys. In J. Cassidy & P. R. Shaver (Eds.), *Handbook of attachment: Theory, research, and clinical applications* (pp. 181–197). New York: Guilford.

Suomi, S. J. (2000a). A biobehavioral perspective on developmental psychopathology: Excessive aggression and serotonergic dysfunction in monkeys. In A. J. Sameroff, M. Lewis, & S. Miller (Eds.), *Handbook of developmental psychopathology* (pp. 237–256). New York: Plenum.

Suomi, S. J. (2000b). Behavioral inhibition and impulsive aggressiveness: Insights from studies with rhesus monkeys. In L. Balter & C. Tamis-Lamode (Eds.), *Child psychology: A handbook of contemporary issues* (pp. 510–525). New York: Taylor & Francis.

Suomi, S. J. (2004). Gene–environment interactions and the neurobiology of social conflict. In J. A. King, C. F. Ferris, & I. Lederhendler (Eds.), *Roots of mental illness in children. Annals of the New York Academy of Science, 1008,* 132–139.

Suomi, S. J. (in press). Using nursery rearing to study gene–environment interactions in rhesus monkeys. In G. P. Sackett & G. C. Ruppenthal (Eds.), *Nursery rearing of nonhuman primates in the 21st century.* New York: Kluwer.

Suomi, S. J., & Levine, S. (1998). Psychobiology of intergenerational effects of trauma: Evidence from animal studies. In Y. Daniele (Ed.), *International handbook of multigenerational legacies of trauma* (pp. 623–637). New York: Plenum.

Suomi, S. J., & Ripp, C. (1983). A history of motherless mother monkey mothering at the University of Wisconsin primate laboratory. In M. Reite & N. Caine (Eds.), *Child abuse: The nonhuman primate data* (pp. 49–77). New York: Liss.

Williamson, D. E., Coleman, K., Bacanu, S. A., Devlin, B. J., Rogers, J., Ryan, N. D., & Cameron, J. L. (2003). Heritability of fearful–anxious endophenotypes in infant rhesus macaques: A preliminary report. *Biological Psychiatry, 53,* 284–291.

Zajicek, K., Higley, J. D., Suomi, S. J., & Linnoila, M. (1997). Rhesus macaques with high CSF 5-HIAA concentrations exhibit early sleep onset. *Psychiatric Research, 77,* 15–25.

9780805895384.